Edward Lhwyd

T0345590

SCIENTISTS OF WALES

Series Editor
Gareth Ffowc Roberts
Bangor University

Editorial Panel
John V. Tucker
Swansea University

Iwan Rhys Morus
Aberystwyth University

SCIENTISTS OF WALES

Edward Lhwyd

c.1660–1709

NATURALIST – ANTIQUARY – PHILOLOGIST

BRYNLEY F. ROBERTS

UNIVERSITY OF WALES PRESS

2022

www.uwp.co.uk

British Library Cataloguing-in-Publication Data
A catalogue record for this book is available from the British Library.

ISBN   978-1-78683-782-0
eISBN 978-1-78683-783-7

The University of Wales Press gratefully acknowledges the support of the
Books Council of Wales in publishing this title.

THE LEARNED SOCIETY OF WALES
CYMDEITHAS DDYSGEDIG CYMRU

CYMYSGEDD
Papur | Yn cefnogi
coedwigaeth gyfrifol
FSC FSC® C013604

Typeset by Marie Doherty
Printed by CPI Antony Rowe, Melksham, United Kingdom

In Memoriam

William Huddesford (1732–1772), Richard Ellis (1865–1928),
R. T. Gunther (1869–1940), Frank V. Emery (1930–1987)

CONTENTS

SERIES EDITOR'S FOREWORD

Wales has a long and important history of contributions to scientific and technological discovery and innovation stretching from the Middle Ages to the present day. From medieval scholars to contemporary scientists and engineers, Welsh individuals have been at the forefront of efforts to understand and control the world around us. For much of Welsh history, science has played a key role in Welsh culture: bards drew on scientific ideas in their poetry; renaissance gentlemen devoted themselves to natural history; the leaders of early Welsh Methodism filled their hymns with scientific references. During the nineteenth century, scientific societies flourished and Wales was transformed by engineering and technology. In the twentieth century the work of Welsh scientists continued to influence developments in their fields.

Much of this exciting and vibrant Welsh scientific history has now disappeared from historical memory. The aim of the Scientists of Wales series is to resurrect the role of science and technology in Welsh history. Its volumes trace the careers and achievements of Welsh investigators, setting their work within their cultural contexts. They demonstrate how scientists and engineers have contributed to the making of modern Wales as well as showing the ways in which Wales has played a crucial role in the emergence of modern science and engineering.

RHAGAIR GOLYGYDD
Y GYFRES

O'r Oesoedd Canol hyd heddiw, mae gan Gymru hanes hir a phwysig o gyfrannu at ddarganfyddiadau a menter gwyddonol a thechnolegol. O'r ysgolheigion cynharaf i wyddonwyr a pheirianwyr cyfoes, mae Cymry wedi bod yn flaenllaw yn yr ymdrech i ddeall a rheoli'r byd o'n cwmpas. Mae gwyddoniaeth wedi chwarae rôl allweddol o fewn diwylliant Cymreig am ran helaeth o hanes Cymru: arferai'r beirdd llys dynnu ar syniadau gwyddonol yn eu barddoniaeth; roedd gan wŷr y Dadeni ddiddordeb brwd yn y gwyddorau naturiol; ac roedd emynau arweinwyr cynnar Methodistiaeth Gymreig yn llawn cyfeiriadau gwyddonol. Blodeuodd cymdeithasau gwyddonol yn ystod y bedwaredd ganrif ar bymtheg, a thrawsffurfiwyd Cymru gan beirianneg a thechnoleg. Ac, yn ogystal, bu gwyddonwyr Cymreig yn ddylanwadol mewn sawl maes gwyddonol a thechnolegol yn yr ugeinfed ganrif.

Mae llawer o'r hanes gwyddonol Cymreig cyffrous yma wedi hen ddiflannu. Amcan cyfres Gwyddonwyr Cymru yw i danlinellu cyfraniad gwyddoniaeth a thechnoleg yn hanes Cymru, â'i chyfrolau'n olrhain gyrfaoedd a champau gwyddonwyr Cymreig gan osod eu gwaith yn ei gyd-destun diwylliannol. Trwy ddangos sut y cyfrannodd gwyddonwyr a pheirianwyr at greu'r Gymru fodern, dadlennir hefyd sut y mae Cymru wedi chwarae rhan hanfodol yn natblygiad gwyddoniaeth a pheirianneg fodern.

PREFACE

My active interest in Edward Lhwyd, first aroused quite fortuitously towards the end of the 1960s when I was working on an unrelated project, was given an opportunity to develop when I was elected to the Sir John Rhŷs Fellowship at Jesus College, Oxford, for 1973–4. I am very grateful to the Principal and Fellows, in particular the late Sir Idris Foster, for their welcome and hospitality at that time as well as for their support. During that year I met and enjoyed the company of the late Frank Emery, Fellow of St Peter's College. With characteristic kindness, he welcomed this tyro into a field that he had long made his own, and not only generously shared with me transcripts, off-prints and references but also gave me a great deal of his time to discuss his research and insights (in addition to regaling me with tales of Welsh rugby history). But for his untimely death in 1987 this book would have been unnecessary. Over the years I have been given a great deal of cooperation and assistance in Oxford by staff at the Bodleian Library, especially at the Weston Library and Duke Humphrey's Library (as it was then called), the Ashmolean Museum and the Museum of the History of Science (Old Ashmolean); the British Library; and the libraries of the universities of Swansea and Aberystwyth. For many years I have been privileged to enjoy the resources and cooperation of the National Library of Wales, Aberystwyth, and 'generations' of its staff have always been unfailingly generous in their assistance.

I was able to complete transcribing most of Lhwyd's correspondence not long after leaving Oxford and, thanks to the skilful efforts of a number of typists, the letters were eventually available in typescript;

the work was financed by the award of a Leverhulme Emeritus Senior Scholarship in 1997, and also a grant from the University of Wales Vernham Hull Memorial Prize Fund. The sheer volume of paper and the magnitude of the task of checking and editing filled me with despair until, quite unexpectedly, Dr Rhodri Lewis suggested including the Lhwyd correspondence in the union catalogue *Early Modern Letters Online* (*EMLO*), created and maintained by the international research project 'Cultures of Knowledge'. Prof. Dafydd Johnston, then Director of the University of Wales Centre for Advanced Welsh and Celtic Studies, Aberystwyth, agreed to collaborate and to adopt the Lhwyd correspondence as part of the Centre's research programme. We were joined by the late Prof. Richard Sharpe, of Oxford University's Faculty of History, and other members of the 'Cultures of Knowledge' project, and together we were able to appoint Helen Watt as research assistant. Her skills as an archivist and historian, her wholehearted enthusiasm for the project and her practical interest in its development were crucial to its successful and timely completion. I owe her a deep personal debt of gratitude.

I am grateful to the editors of the Scientists of Wales series, especially Dr Gareth Roberts, for their kind invitation to participate in the series and to contribute this study. It was the necessary impetus that I required to attempt to impose order on material and notes that I had been collecting for too long a period. Dr Llion Wigley of the University of Wales Press has shown much understanding and sympathy in the inevitable interruptions that occurred during the planning and writing of this book; his forbearance and patience have been a great help. I have frequently profited from the friendship and collaboration of other scholars who have freely given of their expertise and who have always been willing to answer queries, to make suggestions and to draw my attention to relevant publications. Among them are the late Prof. R. Geraint Gruffudd, Dr Dewi Evans, Prof. Nancy Edwards, Andrew Hawke, Dr Frank Horsman, Dr Daniel Huws, Dr Oliver Padel, Dr Colin Thomas, Dr Leigh T. I. Penman, Dr Marion Löffler; Dr Elissa R. Henken, Mary Burdett-Jones and Philip Henry Jones are friends whose practical help and unfailing encouragement have been more crucial than they themselves perhaps realised. I must apologise to those others whose names I have inadvertently not recorded here.

Edward Lhwyd has been my companion for many years, and since 1971 I have published accounts of aspects of his life. The major primary biographical source is his correspondence, comprising over 2,000 letters written or received by him. Almost all of those written by Lhwyd were edited by R. T. Gunther, *Life and Letters of Edward Lhwyd* (1945). These, together with the letters that Lhwyd received, can now be read online and in sequence on the *EMLO* database, *Early Modern Letters Online*, edited by Howard Hotson and Miranda Lewis, at *http://emlo.bodleian. ox.ac.uk*. This provides digital images and transcripts of the letters with translations of those not in English.

Although there are modern studies of many of the scholars associated with Lhwyd, notably John Ray, Martin Lister, William Nicholson, John Aubrey, Thomas Hearne and John Woodward, Lhwyd himself has not been so fortunate. The various attempts at biography, from Huddesford's pioneering 'Memoirs' to Frank Emery's *Edward Lhuyd, F.R.S., 1660–1709* (1971), the best and most balanced account of Lhwyd to date, are discussed in Chapter 11.

Brynley F. Roberts

It has been a privilege and pleasure for Elissa, Mary and myself to prepare Brynley's work for publication. We would like to add to the acknowledgements Arthur O. Chater, Prof. Marged Haycock, Richard Ireland and Gruffudd Antur for responding so speedily to requests for assistance. We thank the editor of the series, Gareth Roberts, and the commissioning editor, Dr Llion Wigley, for their support and understanding during this exceptionally difficult time, and also the staff of the Press for their work.

Philip Henry Jones
October 2021

ABBREVIATIONS

AC	*Archaeologia Cambrensis*
Alum. Oxon.	J. Foster, *Alumni Oxonienses*, 2 vols (Oxford, 1891–2)
ANH	*Archives of Natural History*
Ashm.	Ashmolean Museum manuscript or book. Ashmolean manuscripts are housed in the Bodleian Library, Oxford; Ashmolean books are in the Ashmolean Museum, Oxford
BBCS	*Bulletin of the Board of Celtic Studies*
BL	British Library
Camden's Wales	*Camden's Wales* … with an introductory essay by Gwyn Walters (Carmarthen, 1984)
Considine	John Considine, *Small Dictionaries and Curiosity: Lexicography and Fieldwork in Post-medieval Europe* (Oxford, 2017)
DWB	*Dictionary of Welsh Biography*
ELSH	J. L. Campbell and Derrick Thomson, *Edward Lhuyd in the Scottish Highlands 1699–1700* (Oxford, 1963)
Emery, 'Glamorgan'	F. V. Emery, 'Edward Lhuyd and some of his Glamorgan Correspondents: A View of Gower in the 1690s', *THSC* (1965), 59–114
EMLO	*Early Modern Letters Online*

Feingold	Mordechai Feingold, 'The Mathematical Sciences and New Philosophies', in *The History of the University of Oxford*, vol. IV: *Seventeenth-Century Oxford*, ed. Nicholas Tyacke (Oxford, 1997), pp. 359–448
G	R. T. Gunther, *Early Science in Oxford*, vol. XIV: *Life and Letters of Edward Lhwyd* (Oxford, 1945)
Gunther, *Biological*	R. T. Gunther, *Early Science in Oxford*, vol. III: *Part I, The Biological Sciences; Part II, the Biological Collections* (Oxford, 1925)
Gunther, *Dr. Plot*	R. T. Gunther, *Early Science in Oxford*, vol. XII: *Dr Plot and the Correspondence of the Philosophical Society of Oxford* (Oxford, 1939)
Gunther, *Philosophical*	R. T. Gunther, *Early Science in Oxford*, vol. IV: *The Philosophical Society* (Oxford, 1925)
Hearne	*Remarks and Collections of Thomas Hearne*, ed. C. E. Doble, D. W. Rannie and H. E. Salter, 11 vols (Oxford, 1885–1921)
Huddesford	[William Huddesford], 'Memoirs of the life of Edward Lhwyd, M.A.', in Nicholas Owen, *British Remains, or a Collection of Antiquities Relating to the Britons* (London, 1777), pp. 131–84
JSBNS	*Journal of the Society for the Bibliography of Natural History*
Leoni	Simona Boscani Leoni, 'Queries and Questionnaires. Collecting Local and Popular Knowledge in 17th and 18th Century Europe', in *Wissenchaftsgeschichte und Geschichte des Wissen im Dialog = Connecting Science and Knowledge*, ed. Kaspar von Greyerz, Silvia Flubacher and Philip Senn (Göttingen, 2013), pp. 187–210

Lloyd, 'Correspondence'	Nesta Lloyd, 'The correspondence of Edward Lhwyd and Richard Mostyn', *Flintshire Historical Society Publications*, 25 (1971–2), 32–61.
NLW	Aberystwyth, National Library of Wales
NLWJ	*National Library of Wales Journal*
Ovenell	R. F. Ovenell, *The Ashmolean Museum, 1683–1894* (Oxford, 1986)
Parochialia	Rupert Morris (ed.), *Parochialia Being a Summary of Answers to Parochial Queries in Order to a Geographical Dictionary etc., of Wales issued by Edward Lhuyd. Archaelogia Cambrensis*, Supplements (London: Cambrian Archaeological Association, 1909–11)
Parry	Graham Parry, *The Trophies of Time: English Antiquarians of the Seventeenth Century* (Oxford, 1995)
Pearman	David Pearman, *The Discovery of the Native Flora of Britain and Ireland* (Bristol, 2017)
Piggott, *Ruins*	Stuart Piggott, *Ruins in a Landscape: Essays in Antiquarianism* (Edinburgh, 1976)
Poole	William Poole, *The World Makers: Scientists of the Restoration and the Search for the Origins of the Earth* (Oxford, 2010)
Pryce, *Diocese*	Arthur Ivor Pryce, *The Diocese of Bangor during Three Centuries … being a Digest of the Registers of the Bishops* (Cardiff, 1929)
PT	*Philosophical Transactions*
Raven, *Ray*	Charles E. Raven, *John Ray, Naturalist: His Life and Works* (Cambridge, 1950)

Ray Correspondence	*The Correspondence of John Ray: Consisting of Selections from the Philosophical Letters published by Dr. Derham, and Original Letters ...*, ed. Edwin Lankester (London, 1848)
Repertory	Daniel Huws, *A Repertory of Welsh Manuscripts and Scribes, c.800–c.1800*, 3 vols (Aberystwyth, 2022)
RFC	*Further Correspondence of John Ray*, ed. R. T. Gunther (London, 1928)
Roberts, 'Books'	Brynley F. Roberts, 'Edward Lhwyd's Collection of Printed Books', *The Bodleian Library Record*, 10 (1979), 112–27
Roberts, 'Carmarthenshire'	Brynley F. Roberts, 'Edward Lhwyd in Carmarthenshire', *The Carmarthenshire Antiquary*, 46 (2010), 24–43
Roberts, 'Ceredigion'	Brynley F. Roberts, 'Edward Lhwyd a Cheredigion', *Ceredigion*, 16 (2009), 49–69
Roberts, 'Cymraeg'	Brynley F. Roberts, 'Cymraeg Edward Lhwyd', *Y Traethodydd*, 158 (2003), 211–28
Roberts, 'Folklorist'	Brynley F. Roberts, 'Edward Lhwyd (*c.*1660–1709): Folklorist', *Folklore*, 120 (2009), 36–56
Roberts, 'John Lloyd'	Brynley F. Roberts, 'Llythyrau John Lloyd at Edward Lhuyd', *NLWJ*, 17/1 (1971), 8–114; 17/2 (1971), 183–206
Roberts, 'Memoirs'	Brynley F. Roberts, '"Memoirs of Edward Lhwyd, Antiquary" and Nicholas Owen's *British Remains, 1777*', *NLWJ*, 19/1 (1975), 67–87
Roberts, 'Protégés'	Brynley F. Roberts, 'Edward Lhwyd's Protégés', *THSC*, NS, 14 (2008), 21–57

Roos, *Web*	Anna Marie Roos, *Web of Nature: Martin Lister (1639–1712), the First Arachnologist* (Leiden and Boston, 2011)
SC	*Studia Celtica*
Sharpe, *Letters*	*Roderick O'Flaherty's Letters 1696–1709: To William Molyneux, Edward Lhwyd, and Samuel Molyneux, 1696–1709*, ed. with notes and an introduction by Richard Sharpe (Dublin, 2013)
Simcock	A. V. Simcock, *The Ashmolean Museum and Oxford Science, 1683–1983* (Oxford, 1984)
Texts	Edward Lhwyd, *Archæologia Britannica, Texts & Translations*, ed. Dewi W. Evans and Brynley F. Roberts, Celtic Studies Publications, 10 (Aberystwyth, 2009)
THSC	*Transactions of the Honourable Society of Cymmrodorion*
Uffenbach	*Oxford in 1710 from the Travels of Zacharias Conrad von Uffenbach*, trans. and ed. W. H. Quarrell and W. J. C. Quarrell (Oxford, 1928)
WHR	*Welsh History Review*
ZCP	*Zeitschrift für celtische Philologie*

A NOTE ON DATES, MONEY AND TRANSCRIPTIONS

Julian and Gregorian

Britain kept to the Julian calendar until 1752, when the reformed Gregorian calendar was adopted. This meant that up to 28 February 1700 Britain was ten days behind virtually all western European countries, and eleven days behind from 1 March 1700 onwards. In this work Julian dates have been retained apart from Lhwyd's brief visit to Brittany.

Civil/ecclesiastical/legal year

Until 1751 the official year commenced on 25 March. Thus Queen Elizabeth I who died on 24 March 1603 (by modern reckoning), officially died in 1602. To avoid confusion, dates (often given in the original in the awkward '169¾' format) are based where possible on the modern January to December reckoning.

Currency

Sums expressed in pounds, shillings and pence are reproduced using the abbreviations £ *s. d.*, e.g. £15 16*s.* 6*d.* The abbreviation 'li' employed for pound (often as a superscript) is replaced by the £ character. Sums expressed in Roman numerals have been converted to Arabic.

Value of money

As will become apparent, money (or the absence of it) is a major theme in the story of Lhwyd and his immediate forebears. Various methods have been suggested for converting money to modern values. There is a valuable discussion of the changing value of money in Roderick Floud, *An Economic History of the English Garden* (n.p., 2019), pp. 9–14, where he presents cogent arguments for basing comparisons on average earnings rather than purchasing power as represented by the retail price index. The choice can have dramatic effects; the £4,000 owed by Edward Lloyd to neighbouring gentry in 1674 corresponds to some £618,000 today based on purchasing power but to £7,314,000 based on labour value.

All conversions to modern values employed the calculator available at *https://www.measuringworth.com/calculators/ukcompare/*.

Transcriptions

For the convenience of readers, wherever possible reference is made in the text to the letters as transcribed in R. T. Gunther, *The Life and Letters of Edward Lhwyd*, as (G, pp. xx). Gunther systematically ignored Lhwyd's use of capitals but since Lhwyd had a clear scheme for capitalisation, capitals have been tacitly restored from *EMLO* transcripts and images. On the other hand, EMLO expanded the abbreviations used by Lhwyd and his correspondents. These remain as reproduced by Gunther.

For brevity and consistency, Lhwyd's *Lithophylacii Britannici Ichnographia* is referred to as the *Ichnographia*.

GLOSSARY

Lhwyd and other seventeenth-century collectors of 'figured' or 'formed' stones used a mixture of terms, some colloquial, reflecting popular beliefs about the origin and alleged virtues of the item, others descriptive based on analogy with known life forms.

Asteriae   Star-like fossilised Crinoid (sea lily) columnals (stem segments)

Astroites   Fossil corals with star-like openings, or pentagonal star-shaped stem ossicles of fossil pentacrinites (for pentacrinites, see below)

Belemnites   Solid bullet-shaped fossil guards (internal skeletons) of extinct squid-like cephalopods

Bufonites   Crushing teeth of fossil fish, often of Jurassic genus *Lepidotus*. Traditionally believed to be a panacea found in the heads of toads

Ceraunia   Stones believed to have fallen from the sky, possibly produced by lightning. A very mixed category which included fossils, meteorites and prehistoric implements

*Cochlite*s   Fossil spiral shell

Cock stones or Cock's knee stones   Fossil sea urchin tests, called by Lhwyd *Echinites pileatus minor*

Copperas   see **Pyrite**

Cornu Ammonis   Fossil ammonites, an extinct group of molluscs

Crampstones   Fossilised teeth of sharks; see *Glossopetrae* and **Pyrite**

Crinoids 'Sea lilies', echinoderms; fossil forms were stemmed with five arms radiating from the calyx

Echinites Fossil sea urchin tests

Entrochi Term coined by G. Agricola (1546) for wheel-like fossil crinoid columnals

Fairy beads Small disc-like fossilised columnals of crinoids such as *Actinocrinites*

Fairy saltcellars see *Ichthyospondyli*

Fayrie causeways Term coined by Lhwyd, on a morphological analogy with Giant's Causeway, for polygonal basaltiform corals such as *Lithostrotion*

Glain neidr see *Ovum anguinum*

Glossopetrae Term used since antiquity (Pliny the Elder); their nature as fossilised sharks' teeth, often from the extinct *Otodus megalodon* (formerly *Carcharodon megalodon*), was first recognised by Fabio Colonna in 1616

Ichthyodontes Fossilised fish teeth

Ichthyospondyli Fossilised vertebrae of fish, though Lhwyd had included plesiosaur and ichthyosaur vertebrae in this category in the *Ichnographia*

Lapides Judaici Bulbous fossil spines (radioles) of the Cretaceous echinoid *Balanocidaris glandifera*

Lapides sui generis Literally, 'stones in their own right'; 'formed stones' spontaneously created by nature, hence of non-organic origin

Leechstones see *Siliquastra*

Lithophyta 'stone plants', often corals, but used by Lhwyd in his *Ichnographia* for mineralised leaves and wood

Lithostrotion Term coined by Lhwyd and still in use today for Carboniferous fossil rugose coral

Marcasite   Mineral, Iron sulphide, FeS_2, a form of pyrite, often found as nodules, bronze to silvery polished crystals used as jewels

Ombriae pellucidae   Spherical transparent or translucent pebbles, generally of quartz

Ovum anguinum   Jurassic fossil sea urchin (Cidaroid); its bulbous spines are the *Lapides Judaici*. Popularly believed to be snakes' eggs

Pentacrinites   Jurassic fossil crinoids with pentagonal star-like column

Pyrite   Mineral, Iron sulphide, FeS_2, often called 'fool's gold', brass-yellow cubic or octahedral crystals used in jewellery and historically for the manufacture of sulfuric acid

St Cuthbert's beads   see *Entrochi*

Selenite   Mineral, colourless transparent form of Gypsum, $CaSO_4.2H_2O$, often found as crystals

Siderite   Mineral, Iron carbonate $FeCO_3$, sometimes forming rhombohedral crystals

Siliquastra   Term coined by Lhwyd for fossils resembling the seed pods of lupins or other legumes

Snakestones   see *Cornu Ammonis*

Snakes' eggs   see *Ovum anguinum*

Star-stones   see *Astroites*

Toadstones   see *Bufonites*

Tonguestones   see *Glossopetrae*

Trochites   Term coined by G. Agricola (1546) for isolated circular plates (ossicles) from crinoid columnals; from *trochus*, a wheel or hoop

Zoophytes   Obsolete term for invertebrates resembling plants, such as corals and sponges

LIST OF ILLUSTRATIONS

Apart from the Frontispiece, reproduced courtesy of the photographer, all the illustrations are reproduced courtesy of the National Library of Wales.

INTRODUCTION

Since this study is appearing in the series 'Scientists of Wales', it is appropriate to ask to what extent Edward Lhwyd, who was born and raised in Shropshire and spent most of his working life in Oxford, can be thought of as Welsh and whether he thought of himself as Welsh.[1] One answer, explored in Chapter 1, is that although included in the English county of Shropshire, the Oswestry area remained Welsh in language and culture for much of the seventeenth century.

The genealogy quoted at the beginning of Chapter 1 demonstrates Lhwyd's desire to present himself as a non-English Briton, a descendant of those whom he would elsewhere call the 'First Planters of the Three Kingdoms'.[2] The form of surname he eventually settled upon shows a similar desire to emphasise his Welsh identity.[3] The Llanforda family had adopted Lloyd as the 'official' form of their family name three or more generations before Lhwyd's birth. He matriculated at Jesus College, Oxford, in 1682 as Edward Lloyd and signed himself thus in his earliest surviving correspondence. From about 1688 onwards, his correspondents increasingly addressed him as 'Llwyd', and from 1689 he regularly signed himself 'Edward Lhwyd', an early example of a Welsh-speaker consciously opting to use the original Welsh form of his anglicised surname.[4] When he published his Latin *Ichnographia* in 1699 he employed *Edwardus Luidius*, reflecting a non-Welsh pronunciation of the initial letter but retaining the Welsh diphthong; this is the form that he used regularly in his Latin correspondence. On the title page and Preface of the *Archæologia Britannica* (1707) he used the variant form *Lhuyd*, a spelling which first appears as 'Mr Edwd Lhûyd' in 1694 in the list of members of the Red Herring Club. This form is more in accord with Lhwyd's 'General Alphabet', which uses *û* for the English vowel *oo*. The choice of 'Lhuyd' may have been made to assist non-Welsh readers to pronounce his Welsh name correctly but may also be a tribute to Humphrey Lhuyd of Denbigh (*c*.1527–68), an earlier Welsh scholar of illegitimate birth who had replaced 'Lloyd' by the Welsh original. Lhwyd's own usage was more or less consistent – in Latin, *Luidius*; in the *Archaeologia*, *Lhuyd*; elsewhere, in Welsh and English, *Lhwyd* – but posthumously 'Lhuyd' became the commonest form, probably because it was employed in catalogues and bibliographies.

Lhwyd's closest circle of friends was drawn from his contemporaries at Jesus College, Oxford, and from later cohorts of students there, some of whom became his assistants. Although English (and to a far lesser extent, Latin) was the usual medium of Lhwyd's correspondence, some friends wrote to him in Welsh and he occasionally replied in the same language. Notes and instructions to his assistants were quite often in Welsh, suggesting this was their usual medium of communication.

In his scientific work, comments in Lhwyd's notes show that Welsh was often his language of first response. Marginal annotations in some of his books are in Welsh, some being extended discussions of scientific matters, most notably his scathing criticisms of John Woodward's theories. These represented a conscious attempt to use Welsh for the writing of science and to coin technical terms for such discussions.[5] Lhwyd may have intended to prepare Welsh-language reports for the Welsh gentry and clergy who had responded to his enquiries, many of whom remained fluent Welsh-speakers but had, at best, an ambiguous attitude towards the language. He believed that presenting them with scholarly work in Welsh might develop their pride in the language; he told Richard Mostyn he hoped that Pezron's *Antiquité ... des Celtes* might published in Welsh before appearing in English since 'it would certainly sel very well and contribute much to the preservation of the language amongst the Gentry' (G, p. 492). It was Lhwyd's grasp of the Welsh language that enabled him to develop his innovative theory of the relationship between the various Celtic languages, the greatest and most lasting of his contributions to learning.

For most of his life Lhwyd knew little about contemporary Welsh writing, telling Richard Mostyn in 1707 that his assistant Moses Williams was 'much more conversant than I in printed Welsh' (G, p. 537). He was born a generation too late to have acquired a first-hand knowledge of the Welsh bardic tradition and of the strict metres. His main interest in literary texts was as historical sources or as linguistic evidence; when he remarked in 1698 that the early poets were 'much more worth our acquaintance than is commonly represented' (G, p. 379), he was not thinking of their literary qualities. This did not deter him from venturing to compose a few Welsh strict-metre poems, his first public attempt being a series of *englynion* in the 1695 Oxford University collection of poems on the death of Queen Mary, modelled on the exemplar in Siôn Dafydd Rhys's grammar of 1592.[6]

Wales itself was of vital importance for Lhwyd's fieldwork. Here, at the beginning of his career, he first made his reputation as a botanist by discovering plants that had eluded experienced plant-hunters such as John Ray and Francis Willughby. Here, too, he encountered far older rocks than the Mesozoic and Cenozoic strata he had explored in the Oxford area, and found important fossils, including the first trilobite recorded in Wales. Even Welsh minerals could become sources of patriotic pride: it was probably no accident that the first specimen listed in his *Ichnographia* was an impressive quartz crystal, *Crystallus maxima Britannia*, which he himself had found *in Alpibus Arvoniæ, iuxta lacum Fynon Vrech* (in the alps of Arfon, next to Ffynon Frech), that is, in Cwm Glas on Snowdon, the site of some of his most important botanical discoveries. Similarly, Welsh mountains were exceptionally rich in rare plants: 'the Mountains of Llan Llechid and Llan Beris in Carnarvanshire, afford more sorts of *Alpine* plants, than have been as yet discover'd on all the other mountaines of the Isle of *Britain*.'[7] Sometimes Lhwyd and his correspondents employed the Welsh names of plants and animals, such as *chwain y môr* (sea fleas). These are often the earliest attestations of these words in *Geiriadur Prifysgol Cymru* and are of particular value since they derive from contemporary usage rather than dictionary definitions.

Lhwyd's pride in Wales is particularly evident in his additions to the 1695 revision of Camden's *Britannia*. Responding to Camden's disparaging claim that Merioneth was 'the roughest and most unpleasant County in all Wales', Lhwyd proudly replied that if 'variety of subjects make a Country appear delightful, this may contend with most; as affording (besides a sea-prospect) not only exceeding high mountains, and inaccessible rocks; with an incredible number of rivers, cataracts, and lakes: but also variety of lower hills, woods, and plains, and some fruitful valleys.'[8] His patriotism is remarkably similar to that of his contemporary, the Swiss naturalist J. J. Scheuchzer, who expressed his pride in his own 'sweet fatherland', which was 'not harsh and wild' but possessed 'so many and such great beauties and such heart-warming gifts of Nature'.[9]

As Elizabeth Yale has recently pointed out, Lhwyd was active at a time when 'national distinctions' and the relationships between England and its three peripheral countries were a live issue;[10] in 1707, the year the first volume of Lhwyd's *Archæologia* was published, extensive bribery had persuaded the Scottish parliament to take the very unpopular step of dissolving itself. When reviewing the book, William Baxter found it necessary to defend Lhwyd's impartiality against those who claimed it was 'design'd to serve a certain Interest'. In fact, those who believed that Lhwyd was constructing an alternative narrative of British history by demonstrating the shared identity of the ancient Britons, had indeed understood the implications of his work.

A modern scientist might ask to what extent Lhwyd might be considered a scientist as he moved from botany, mineralogy and palaeontology, all of which would be considered 'science' today, to field archaeology, epigraphy and linguistic studies designed to cast light on historical questions. There are two responses, the first being that linguistics, as its German name *Sprachwissenschaft* indicates, is a science. The second response is that the question is anachronistic. Since modern science is largely based upon a mathematical, statistical and probabilistic approach to the data, there has been a tendency to emphasise those considered to be its forerunners, such as Newton, Halley and other 'geometers'.[11] In fact, though very influential, they comprised a small minority, being greatly outnumbered by the naturalists, many of whom, like Lhwyd himself, did not find it necessary to adopt a mathematical approach.

Seventeenth-century scholarship viewed the whole of creation and its history as its field of study. There was no sharp dividing line between the natural and the human world; the search for origins and relationships provided a conceptual basis for the collecting, description, ordering and interpretation of objects that aroused the curiosity and occupied the energies of so many scholars. To Lhwyd's contemporaries, natural history and the study of antiquities went hand in hand. Whatever the nature of the evidence, be it physical, written or oral, what mattered was adopting a scientific approach: careful observation, preferably in person or by a reliable witness; meticulous description; systematic organisation of evidence; separation where possible of description and explanation, as in the *Ichnographia*, and, for Lhwyd, the rejection of non-materialistic, 'praeternatural' causes. Those of his explanations which contemporaries found hardest to accept – that the mysterious Merioneth fires had been caused by rotting locusts, or that fossils grew from spores washed down into the rocks – were those he had been compelled to offer to avoid supernatural causes such as witchcraft for the former[12] or some 'plastic' force or 'virtue' for the latter. As a true scientist, he recognised the provisional nature of his hypotheses: 'for they who have no other aim than the search of Truth, are no ways concernd for the honour of their opinions' (G, p. 393). Unlike present-day scientists, he accepted final causes, 'there is an End in all the productions of Nature', but greatly weakened their explanatory value by arguing that 'we are often but very improper judges of such final causes' (G, p. 396).

Readers will find that what follows is a male-dominated narrative, since in Lhwyd's day science was pursued by gentlemen, men who had been born wealthy or had received a university education. As will be shown in Chapter 3, gentlemanly conventions governed the rituals of conducting a scholarly correspondence with its reciprocal exchange of information, specimens and sometimes other gifts; in 1697 Ray thanked Lhwyd for 'a large Cheshire cheese ... as good as it is great'.[13] Gentlemanly norms also underlay the frequent references in letters to drinking healths *in absentia* and to arranging convivial meetings; thus Edmund Gibson told Lhwyd in February 1694, 'on Saturday come se'night in the evening I may desire your company over a glass of ale at Tom Swifts.'[14] Those who did not enjoy these social or educational advantages were looked down upon, Lhwyd describing the collection of William Cole as being 'very extraordinary ... for one person ... who, perhaps, had not the advantage of a liberal education to invite him to such studies' (G, p. 104). Much of Lhwyd's fossil collecting relied upon paid workmen, whom he viewed with contemptuous suspicion: 'these countrey fellows I find will sell their best customers for a half-penny' (G, pp. 280–1). He was, however, ready to gather information from his social inferiors; an illiterate shepherd could be a trustworthy source for place-names, and colliers knew about 'coal-plants' long before naturalists noticed them.

1

THE LLOYDS OF LLANFORDA AND LHWYD'S GRANDFATHER, EDWARD LLOYD

> I don't profess to be an *Englishman*, but an old *Briton* ...
> descended in the Male-Line from *Heliodore Leathanuin*, the
> Son of *Mercian*; the Son of *Keneu*, the Son of *Coel Killsheavick*,
> (alias *Coel Godebog*), in the Province of *Reged* in *Scotland*, in the
> Fourth Century, before the *Saxons* came into *Great-Britain*.[1]

Edward Lhwyd shared the fondness of Welsh gentry for setting out their honourable and ancient (even legendary) descent and took pride in his detailed knowledge of his lineage. His father claimed in March 1681, perhaps with some exaggeration since it was in an open letter to prospective employers who might wish to have their pedigrees organised, that Edward 'eates drinks & sleepes pedigrees'.[2] Over twenty years later, Lhwyd traced some forty generations of the Lloyds of Llanforda from his father back to the legendary 'Coel Godebog'.[3] He had learned this at home, where his lineage had been preserved in manuscripts compiled by bards and heralds and used by poets in singing the praises of their patrons. Lhwyd's references to his Lloyd descent appeared late in his career and may reflect his developing interest in it at that time. Thus in 1707, when sending Humphrey Foulkes (1673–1737) a list of the Fifteen Tribes of Gwynedd he could not resist adding, after the name of Hedd Molwynog, 'whence the Lhwyds of Havod unos, Lhwyn y Maen, Lhanvorda, Drenewydd, Blaen y Ddôl &c. 1170.

He was descended from Lhywarch hen and so to Koel Godebog' (G, p. 540). The recollection of such a venerable lineage may have been particularly important for Lhwyd, as an illegitimate son born into a family which had recently experienced financial ruin. Despite his illegitimacy, Lhwyd's family on his mother's side, the Pryses of Gogerddan in north Cardiganshire, would prove far more valuable socially and professionally to him than the Llanforda descent. His recognition as a member of the influential Gogerddan family provided him with a range of connections who could assist him as his academic work developed. This offset the ruin of the Llanforda estate and the loss of reputation of the Lloyd family. Nevertheless, it was Llanforda, not Gogerddan, which played a crucial role in fostering the development of Lhwyd's interests.

Both the Lloyds of Llanforda, a township on the high ground about a mile and half west of Oswestry, and the Lloyds of the neighbouring Llwyn-y-maen estate traced their descent from Meurig Llwyd, lord of Henllys, Llangernyw, Denbighshire.[4] According to the imaginative family history, Meurig had been obliged to flee,

> when he hanged those who had come there bringing English laws and customs ... Because of this he forfeited his land to the king ... Because of his valour and bravery John fitz Alan ... took him under his protection and made him Captain General in the siege of Acre [1191] ... in which battle he won his arms which his descendants assert to this day: viz. a two-headed eagle on a field argent without distinction between it and that of the emperor except for the colour of the field. And the reason that the Herald King of Arms granted him these arms was that he regained the standard and banner of the emperor from his enemies.[5] (translated)

Meurig Llwyd found favour with Ieuan Fychan, Constable of Knockin, whose daughter and heiress Agnes he married; she brought with her the estates of Llwyn-y-maen and Llanforda. Richard Lloyd, fifth in line from Meurig Llwyd, died in 1508, leaving Llwyn-y-maen to his second son Edward and Llanforda to his elder son, John.

Although their estate, located in the centre and western part of Llanforda township, was not large, the Lloyds of Llanforda, became one of the leading families of Oswestry and its neighbourhood. Following the Reformation both branches of the Lloyd family became ardent recusants. A 'horseload of papist books, and crucifixes' was seized at Llwyn-y-maen in 1597, and Edward Lloyd (*c*.1570–?1648) was named as 'a notorious recusant' in 1606.[6] An extremely quarrelsome man, he frequently found himself in court.[7] While imprisoned in London in 1619, he inflamed Parliament by his intemperate criticism of Frederick, Elector Palatine, and his wife Elizabeth, idolised by Protestants as victims of Catholic aggression. He was fortunate to escape severe penalties, and subsequently lapsed into obscurity.[8] A tendency to disproportionate verbal aggression surfaced from time to time in the Lloyd family, notably in Edward Lloyd, Lhwyd's father, and, possibly, in Lhwyd himself, as in his acrimonious dispute with Woodward discussed in Chapter 7. John Lloyd, the elder, of Llanforda was also named as a recusant and excommunicated in 1607,[9] but as time passed the Llanforda branch of the family abandoned the old faith and became devout high Anglicans. The estate was extended by several small purchases in the early seventeenth century, and the status of the family, connected by marriage and descent with many marcher families in Shropshire and Wales, increased. The family proudly displayed the double eagle of their heraldic arms on the font in Oswestry church with the date 1662, on the family burial vault and on the south front of the imposing half-timbered edifice, known as the Lloyd Mansion in Cross Street, Oswestry, a late medieval building remodelled around 1600 by a John Lloyd, perhaps as a residence for one of the family.[10]

Oswestry, though not as strategically important as its larger neighbour Chester, controlled access to the fertile lowland of north Wales and the route to Ireland. Its castle and town walls bore witness to its importance throughout the Anglo-Welsh conflicts of the twelfth and thirteenth centuries, while the granting of its charter in 1263 and its Royal Charter of 1398 testified to its growing commercial importance. Oswestry's markets and shops served communities on both sides of a border that was not fixed until the Acts of Union of 1536–43, making it a place where Welsh and English mingled freely to such an extent that 'by the late Middle Ages Oswestry was quintessentially a Welsh town.'[11] It was more Welsh in character than the English boroughs planted in Wales, its population of gentry families, burgesses and civic officers, merchants and tradesmen being kin to many families in north and mid-Wales. Indeed, Oswestry and its western townships of Llanforda, Treflach, Blodwel, Trefonen and Llynclys, were probably regarded by the Welsh gentry as part of Powys, where many of them had their origins, rather than of Shropshire. The Welshness of Oswestry was further reinforced by its belonging to the diocese of St Asaph until the disestablishment of the Church in Wales in 1920, when it was transferred to the diocese of Lichfield. The contrast between the warmth of Welsh feeling for Oswestry and Welsh hostility towards Chester is exemplified in a poem by Lewis Glyn Cothi (*fl.* 1447–86) in which he requests the gift of a sword from Dafydd ap Gutun of Oswestry to enable him to avenge the mistreatment he had received in Chester.[12]

In Oswestry and in the gentry houses of its outlying townships, traditional Welsh culture flourished, maintained by the same structure of patronage and bardic praise as in Wales itself. As well as eulogising the town, some of the leading Welsh poets of the fifteenth and sixteenth centuries made their homes there. Guto'r Glyn (*fl.* 1440–93) and Tudur Aled (*fl.* 1480–1526), both honorary burgesses, praised Oswestry's buildings and institutions and its patronage of poets.[13] When praising Richard Lloyd of Llwyn-y-maen, Tudur Aled called himself 'his Taliesin', that is, the archetypal court poet.[14]

In the sixteenth century the poet Wiliam Llŷn (*c.*1534–80) numbered Oswestry gentry, including the Lloyds of Llanforda, among his patrons.[15] He left his house in Willow Street to his widow Elizabeth for her lifetime;[16] in 1587 she let it to Llŷn's pupil, Rhys Cain (*c.*1546–1614), to whom Llŷn had bequeathed his 'books', the manuscript collections of poems, genealogies and pedigree rolls which enabled Cain to inherit his teacher's patrons, including John Lloyd of Llanforda, whose elegy he composed in 1603.[17] These manuscripts were inherited by his son Siôn Cain (*c.*1575–1650), by whose day social change and Anglicisation had brought about the rapid disintegration of the Welsh bardic order. Although he continued to make a circuit of his patrons, commemorating them and their families in his poems, Siôn realised that producing coats of arms and pedigrees in English had become more profitable than versifying genealogies in Welsh elegies which were largely unintelligible to younger patrons.[18] He was recognised as one of the last representatives of a dying order,[19] and after his death genealogy became the preserve of gentleman antiquaries.

By the mid-seventeenth century the Welsh gentry around Oswestry had become bilingual and bi-cultural, combining the lifestyle and social responsibilities of their class in Wales with the interests and culture of contemporary English gentry society, sending their sons to grammar school and then to university or the Inns of Court. John Lloyd married Mary Lettice (1585–1656), daughter of George Caulfeild MP (1545–1603), Second Justice of the Brecknock circuit from 1601 to his death.[20] Of their nine children, twin sons, Edward and John, both baptised 22 August 1609,[21] lived to survive their father. Both were sent to Oxford, matriculating at Lincoln College, 16 March 1626/7, Edward graduating BA on 24 October 1628, and John on 15 December 1629; John took his MA on 4 July 1633.[22] When Lloyd died in March 1634[23] Siôn Cain composed a traditional elegy.[24] In his will Lloyd left £50 to his wife, and £100 each to his sons, all his books to be shared equally between them. John did not enjoy his inheritance for long, dying in December 1635. His brother Edward, the grandfather of Edward Lhwyd, married Frances, daughter of John Trevor of Bryncunallt (1602–c.1643),[25] who bore him a daughter and two sons (see Chapter 2). On inheriting the estate he began to develop Llanforda gardens on a very ambitious scale, his expenditure alarming his mother.[26] The outbreak of the Civil War brought about a drastic deterioration in his finances and those of his descendants.

Like most Shropshire gentry, Edward Lloyd rallied to the Royalist cause when Charles I came to Shrewsbury in September 1642 to raise money and recruit troops for his intended march on London. Lloyd's 'actions & sufferings in & for his Matie's service' are briefly outlined in the 'true Narrative', an account, written by his son, Edward (Lhwyd's father), in 1662.[27] A recent study of the war in Shropshire provides a broader context for this account.[28] Since Lloyd, like many of the gentry, would have had little ready money at his disposal, he 'gave his Majesty all his plate', which was turned into coinage to pay the Royalist army at the mint which Thomas Bushell had moved from Aberystwyth to Shrewsbury in October 1642.[29] Lloyd then 'raysed a Troope of Dragoons, & armed them on his owne charge, advancing likewise a month's pay to eury soldier out of his owne purse, & served w'th them in p'son vnder the Com'and off Sr Robert Howard ... which Com'ision bore date August the 20th, 1642'.[30] This was further proof of Lloyd's Royalist zeal, since Howard's regiment was probably the first to be raised in Shropshire.[31] It is not known whether Lloyd's dragoons served at the battle of Edgehill on 23 October 1642, and he probably parted company with them when Howard's regiment joined the king at Oxford by the end of the year. The 'true Narrative' continues: 'then being desirous to embrace all opportunitie yt might promote his Matie's service, he proceeded to rayse a Troope of horse vnder the lord Capell, & served with them.' Since the cost of equipping and maintaining a cavalryman was far greater than for a dragoon, this represented a considerably heavier investment by Lloyd in the Royalist war effort. Capel's appointment as Lieutenant-General of Shropshire, Cheshire and North Wales on 4 April 1643 provides a date for the start of Lloyd's command.[32] Oswestry was fortified and garrisoned for the king with Edward Lloyd as its Governor. To prevent the church, which stood outside the town walls, being used as a vantage point by Parliamentarian forces, the famous steeple was reduced to the level of the nave and chancel, and many adjacent houses (some of them Lloyd's property) were demolished.

Lloyd's governorship gained him a place in local folklore. The story, first reported in 1644,[33] was that colonel Thomas Mytton (1596/7-1656), a vigorous Parliamentarian leader, was well aware of Lloyd's love of food and drink. Lloyd was to be invited to dine at a neighbouring house, 'and after he had indulged sufficiently in the pleasures of the table, a party of soldiers from the Garrison of Wem, were to enter the dining room, and make him prisoner, and then to possess themselves of the Castle of Oswestry.' When Mytton's scouts were intercepted, Lloyd was warned and managed to escape. Whatever the truth of the story, it is worth noting that Sir Thomas Eyton at Buildwas later succumbed to a similar plan.[34] Lloyd's removal from command at Oswestry was part of a broader pattern, as Prince Rupert replaced the local military governors of Ludlow, Oswestry, Bridgnorth and Shrewsbury with experienced veteran soldiers.[35] A Royalist newsbook suggests that Lloyd resented being removed from his post: 'upon the Prince declaring Sir *Abram Shipman* Governor of Osestery, Colonell *Loyde* and all his forces quited the Towne', and in their absence the town would have declared for Parliament had not a Royalist force hastened there.[36] Mytton's spies and 'intelligencers' had made him well aware of Oswestry's weakened garrison and of Shipman's temporary absence before he attacked the town on 22 June 1643.[37] After the church and town had been captured, the Royalist garrison retreated into the castle. Early the next day a Parliamentarian officer met 'a party of women of all sorts down on their knees confounding him with their Welch howlings', a testimony to the strength of the language in Oswestry. When an interpreter was found, the women begged to be allowed to persuade the garrison to surrender.[38]

As a result of Capel's many failures, he 'surrendered his Com·mand to ye Lord Byron & his Regiment off horse to Col. Trevor'. John Byron (1598/9–1652) was sent by Prince Rupert in November 1643 to reconquer Lancashire, and by December had reached Chester, where he was given the rank of field-marshal in Lancashire, Cheshire and the six counties of north Wales.[39] Lloyd's kinsman Marcus (or Mark) Trevor (1618–70) took over Capel's regiment in late 1643 or early 1644,[40] and Lloyd accompanied Trevor to Yorkshire and the defeat at Marston Moor 2 July 1644. Hastening back to Wales, he participated in another Royalist defeat at Montgomery on 18 September. Despite outnumbering their opponents, most of the Royalist foot were captured, and the Royalist cavalry, after the initial success of Trevor's horse, 'ran shamefully when they had no cause of fear'.[41] Following these 'fatal Battailes' Lloyd accompanied Trevor to Rhuthun castle, where on 19 October Trevor and his horsemen fled when attacked by Sir Thomas Myddelton, an inglorious incident omitted from the 'true Narrative'. Lloyd was then ordered to south-west England where he served 'till the dismal disbanding at Truro March ye 20th 1645'. In a declaration preceding negotiations for the Royalist surrender, Sir Thomas Fairfax, the Parliamentarian General, stated that upon surrender and a promise never again to bear arms against the Parliament, 'Gentlemen of considerable estates' would receive safe conduct and his 'recommendation to Parliament for their moderate composition'.[42]

Back in Llanforda by May 1645, a temporarily chastened Lloyd faced financial ruin. His houses, kilns and malting rooms in Oswestry, worth a thousand pounds, had, he claimed, been maliciously burnt by the enemy, and his personal estate amounting to six hundred pounds had been plundered. Both sides had resorted to sequestration of enemy estates (seizure of their income) to finance their armies, and Llanforda was now subject to these Parliamentary impositions. Writing to his 'euer honored mother' on 1 May 1645 from his 'ruined home', 'sequestred Llanvorda', he asked 'wheir shall I now serue ye Lord & kneele unto ye Lord my maker?'[43] Oswestry church (twice the scene of hand-to-hand fighting in 1644) lay in ruins.[44] Lloyd described it as being 'like Gods house at Jerusalem, not one stone on another, hard ffortune!' Since he would not worship in a 'conventicle' with 'all sorts of heretickes ... the most of them are Atheists & ye best of the Jewes', he would resort to his house and gardens: 'I will prayse him in the one & pray unto him in the other.' His garden was a place for 'divine meditations ... all howrs of ye day'.[45] There follows a remarkable passage which may help to explain his son's devotion to gardening:

[There] is nothing that growes, but a divine & an holie use may be made on it. Nor hearbe so meane, but speakes the great power off the Almightie: & preaches without treason ... divine Contempla:tion can make gardens as pleasing to the soule as they are delightfull to ye bodie ... I haue been charged wth folly for my gardens & walkes, for my wildernes & ffontaine, & you deare mother haue been at a distane wth me for them as to ye Charges ... but off all transitories their is not such a noble & gentlemanlike vanitie I may say Christian like : as gardens & walkes.

And Deare mother when Man was in ye state off innocence, their was noe place good enough for him but Eden a Garden. &

when your son contemplates heaven he cannot better doe it than in a garden ... Love a garden I beseech you.

He concludes the letter: 'your son having past by, his looser dayes, & first freedom & (as thay say) spent his wild oates, may yet becom graue & settled.'

Lloyd's attempts to continue the education of his son Edward in Thomas Chaloner's school in Overton in 1647 showed that defeat had not blunted his Royalist zeal. Thomas Chaloner MA (*c*.1600–64), a graduate of Jesus College, Cambridge, had been appointed headmaster of Shrewsbury Free Grammar School in 1636.[46] An ardent Royalist, in 1642 he 'lent' the King £600 from the school's chest, for which 'misappropriation' he and others were sued in 1650. Following the Parliamentarian capture of Shrewsbury in February 1645, Chaloner was ejected and commenced a peripatetic existence as a schoolmaster in Shropshire, Staffordshire, Flintshire and Denbighshire. For some nineteen months from July 1647 onwards he kept school at Overton, Flintshire. As well as his former pupils from Hawarden, what Chaloner described in his diary as an 'incredible multitude of gentlemen's sons' were attracted there, Lloyd's son, Edward, being one of them. Writing as one dispossessed Royalist to another, Lloyd praised Chaloner for inculcating sound principles: 'You teach children obedience to ye Lords anointed, & so grafft it vnto them yt thay growing men may become vnderstanding Royalists wch is Treason in this trayterous age.' Lloyd also appreciated Chaloner's teaching skills: 'The way you have of Cozeninge children from their prettie fooleries in to Learning', and enquired where he had re-established his school, for 'Ned', then aged about twelve, 'hath not since Gods wrath parted him & you ... been at any schoole I conceiving it to be ye same thinge to be out of schoole as to be in any but yours'.[47]

His son's education was further disrupted by the Second Civil War. Although Lloyd had told his mother he might 'become graue & settled', he could not resist the call to arms and 'received from ye Lord Byron' a commission as colonel 'to rayse & com'and a Regiment of horse' in Shropshire.[48] Byron, sent from France in the spring of 1648 to coordinate Royalist risings in north Wales, Cheshire and Shropshire to clear the way for the invading Scottish army, found the Welsh gentry were unwilling to serve under him, partly because of his open contempt for their military ability.[49] The Royalists found little support in north Wales apart from a rising in Caernarfonshire headed by Sir John Owen of Clenennau (1600–66), crushed on 5 June 1648 at Y Dalar Hir near Llandygái, and a rebellion in Anglesey, which was defeated at Red Hill near Beaumaris on 1 October. Lloyd's kinsman, Colonel Richard Lloyd of Llwyn-y-maen, was captured at Y Dalar Hir, as were several other Shropshire Royalists, which raises the possibility that Edward Lloyd was also present but managed to escape.[50] If so, whether he, like other Royalist fugitives, subsequently managed to join the rebels in Anglesey remains an unanswered question.

Following the Second Civil War Lloyd was a fugitive for several months. He wrote to his wife from Gogerddan, where his kinsman, Sir Richard Pryse, had promised his support. From there he went to Machynlleth and on to Clenennau in Merionethshire, where he was received kindly,[51] before going on to 'Lleine' (presumably Llŷn), where another kinsman, 'a man of integritie', received his party and their horses 'very civilly'. Following his safe return home he wrote 10 September 1649 to 'ye only man of God yt officiates in llheine', his second cousin, Cadwaladr Kyffin.[52] Lloyd expressed his gratitude by presenting Kyffin with 'your coate of Armes, the last thing, our British Authentique Herald Cain did passe his approbation, before he quit the stage'.[53]

The 'true Narrative' continues by noting Lloyd's 'forwardness to rise w'th Sr Thom: Harries & others', the plot by Sir Thomas Harris of Boreatton and Ralph Kynaston of Llansanffraid-ym-Mechain to seize Shrewsbury castle and Chirk castle on 8 March 1655, one of the many plots betrayed by the Parliamentarian double agent Sir Richard Willys. The final (and perhaps most hazardous) item in Lloyd's list of Royalist fiascos was that 'in Sr George Booths Busines he raysed a Troope off horse (he himselfe beeing then prisoner in Salop) w'ch he put his eldest son to Com'and, And that had proved his and his sonns final ruine, iff his Matie, had not been, by Gods extraordinarie mercy, soe miraculou-slie restored.'[54] This was the premature attempt by Sir George Booth

(1622–84) in August 1659 to restore the monarchy. As usual, the plot had been betrayed to the authorities and was further weakened by poor coordination. In his account of expenses incurred in the Royalist cause, Lloyd noted: 'expended & suffered in & for Sr Georg Booths busines ... betwixt him & his sonn forced to flie into Ireland to secure himselfe £400 0 0' (£57,000 in present-day purchasing power).[55]

The 'true Narrative' speaks of his 'frequent imprisonments, at Least a dozen times' and makes particular reference to his 'beeing fetchd from his house to ye gatehouse, Westminster,[56] for designing the killing of Bradshaw[57] for murthering of his master'. Nor was his family spared: 'his wife was likewise questiond for her life & hardlie escaped, a friend of hers hanged & a maid servant whipt to death for their fidelitie to his Matie, in endevoring to restore the garrison off Oswistree to their pristine Loyalty.' His wife's ordeal was commemorated on her memorial in Oswestry church, dated 15 December 1661: 'Who bore her sexe with peril of her life, / A loyal subject and a loving wife'.[58]

Edward Lloyd suffered severe financial losses both during and after the war. Llanforda was sequestrated for five years. Unless pardoned, Royalists had to compound with Parliament for their 'delinquency', and to recover their estates had to pay severe fines ranging from one sixth to a half of the estate's value. Many, like Lloyd, had to mortgage or sell some of their land, to the detriment of the long-term survival of their estate because of the consequent reduction in income. On 22 July 1646 Lloyd petitioned to be 'admitted in the Truro Articles' and embarked on a long campaign to argue the case for reducing his penalty with the Committee for Compounding. On 24 May 1649 his fine was assessed at £224, subsequently rebated to £200 (some £27,000 in modern purchasing power).[59] Later, he also had to pay £57 (over £10,800 in modern purchasing power) in decimation, a special tax on the estates of notorious Royalists introduced following the risings of 1655 to cover the costs of a Parliamentary militia.[60]

Uncertainty about the fate of the sum settled on his widowed mother added to Lloyd's financial anxieties. He was appalled to learn of her intention to marry Richard Lloyd of Llwyn-y-maen. His stylishly written letters to her express a genuine disgust, as in his letter of

20 October 1648: 'Be not deceived, it is not a noble heate but a depraved itch yt inclines one of your age to marriage.' He was also anxious to protect his patrimony, pointing out bluntly that she was wanted not for her 'burdensome body', but for her wealth and plentiful fortune.[61] There may have been some grounds for his fears. The Committee for the Advance of Money had examined the case of Richard Lloyd in May 1649 and imposed a fine at one-sixth, £480, which had been paid by February 1650. Now, however, the Committee noted that 'by marriage with Lettice Lloyd widow he has got a good personal estate of £160 a year', which they seized until satisfied that he had 'no interest or expectancy of the estate before he compounded'. This was the 'no mean fortune', over £30,000 a year in modern purchasing power, that Edward Lloyd had earlier referred to in a letter to his mother. The matter was settled in October 1651, and Richard Lloyd's case was discharged. Despite her son's protestations, Lettice married Richard Lloyd, dying a few years later on 3 February 1656.[62]

The Llanforda estate never recovered from the financial problems caused by Lloyd's unyielding support for the king. According to the 'true Narrative', his outlay and fines amounted to £8,080 14s. 6d., in modern purchasing power terms £1,222,000. This figure may be inflated, but raising two troops of horse and employing a Captain of Horse for three years would have been a heavy drain on resources, and sequestration for five years, at a cost estimated by Lloyd of at least £2,000, would have contributed to the ruin of the estate, already reduced in value by 'the waste land' caused by the destruction of the houses around Oswestry church. Like many Royalist gentry, Lloyd withdrew from public life and in his 'ruined home' recalled past pleasures; he particularly missed the great delight he had enjoyed 'in our national and British musicke; the house called Llanvorda ... was never without this Hierogliphicke of our Country & Nation', and he sought to borrow a harp in place of his 'sequestered & silenced Jrish and wealsh Harpes'.[63] A chastened man, he sought solace in his gardens.

Although Edward Lloyd lived to see Charles II restored in 1660, he received no compensation for his losses or honours for his efforts. He died 13 February 1663 and was buried 3 March[64] in the family vault in

St Oswald's church, Oswestry, his epitaph commemorating his loyalty: 'One who durst be loyal just and wise / When all were out of countenance here lyes'. He was the first Lloyd for several generations not to be the subject of a Welsh elegy.

2

LIFE AT LLANFORDA: FATHER AND SON

Three children were born to Edward Lhwyd's grandfather, Edward Lloyd, and his wife Frances: a daughter, Anne, baptised 20 March 1634,[1] who married Thomas Kyffin of Sweeney, and two sons. The younger, George, baptised 16 April 1639,[2] entered the Inner Temple 22 October 1656[3] but died in 1659. The elder son, Edward, was baptised 19 July 1635[4] and, as we have seen, was sent to Chaloner's school in Overton. The Second Civil War disrupted his schooling, and he does not appear to have attended university in England. During the early 1650s he may have spent some time on the continent, possibly even attending university there, perhaps in Paris. One of his letter books contains pages of *Sententiae … ex Seneca philosophi collectae*, and *Sententiae ex Satyri: Euphorionis Lusinimi collectae*,[5] while NLW, Peniarth MS 357B bears on its final page the signature of 'Edward Lloyd Lhanvorda Armiger' and contains extracts of some of Aristotle's works in Latin, transcripts of letters in French, Latin poems and French alchemical material. Lloyd may have returned to Llanforda by the mid-1650s and was certainly in Shropshire in August 1659 when he participated in the Booth Rebellion and subsequently fled to Ireland. By the Restoration, Lloyd's father was in poor health and had an uneasy relationship with his son. On 2 September 1662 Edward Lloyd, junior, writing from London, complained that 'my father hathe taken away that smalle allowance w'ch hee was pleased formerly to afford mee for a liuelyhood, & doth nott allowed mee anything at all towards a subsistance',[6] and it was probably then that he was compelled to pawn his cloak.[7]

This sanction may reflect his father's displeasure at the arrival of an illegitimate grandson, the future Edward Lhwyd. Around 1660, Lloyd entered into a relationship with his young kinswoman, Bridget Pryse, the daughter of Thomas Pryse of Glanfrêd, in the parish of Llanfihangel Genau'r-glyn in north Cardiganshire, a nephew of Sir Richard Pryse of Gogerddan. What ensued was a tale of her infatuation with Lloyd, an unwanted pregnancy, a failed attempt to induce an abortion, and her despair.[8] Rejected by her father, Bridget begged Lloyd to marry her, or at least to give her some financial support, but to no avail, since Lloyd realised that without her father's blessing Bridget had no prospects. The year of Lhwyd's birth remains a mystery. When he was first entered in the Jesus College buttery books,[9] in October 1682, his age was given as 18, giving a birth date of *c*.1664. However, when Lhwyd died in 1709 he was said to be forty-nine and so would have been born in 1659–60. Lloyd's will suggests that Lhwyd was a minor (under twenty-one), when Lloyd died in July 1681. If so he would have come of age the following September and would thus have been born *c*.1660. A note, written by Lhwyd himself, adds some information: 'Mr Edward Lhwyd was born at Lappiton parish, his nurse is now living at Krew green. (where he was nurs'd) ... his nurse says he is 41 years old 3 days before Michaelmas last according to Catherine Bowen, his nurse to ye best of her memory.'[10] This was probably written while Lhwyd was in Montgomeryshire in October–November 1698. Since Michaelmas is 29 September, Lhwyd's birthday, '3 days before michaelmas last', would have been 26 September. Unfortunately Catherine Bowen was more sure of Lhwyd's birthday than his birth year, and in the absence of any firm record, '*circa* 1660' is a reasonable surmise for the year of his birth.

Although 'Neddy' was well cared for by his foster parents, after visiting him Bridget told Lloyd that he needed intellectual stimulation: 'Truely yr child though I say it is witty enough and yt would be improved in him dayly if hee weare but to be Instructed and talked to by Ingenious persons.'[11] It would be better for him to be brought up in Llanforda. She also hoped that Lloyd would give her 'something to Maintaine my selfe yt may make me live at some Content', since her father was 'very backward to doe me a kindnesse in this nature', reproaching her that 'I have been an expence to him occation'd by my friendshippe with you.' Lloyd agreed that Neddy should live at Llanforda, though the relationship would not be an easy one. Later, after Bridget's father had relented and allowed her to return to Glanfrêd in the mid-1660s, Neddy stayed with her, but how often and how long these visits were is not known. Bridget and Lloyd remained in regular correspondence, possibly visiting each other up to Lloyd's death,[12] and she participated in his fishing venture described below.[13]

On his father's death in 1663, Lloyd inherited his debts and unpaid fines. Since relatives ignored his pleas for assistance he had to turn to the neighbouring gentry; as one historian has remarked, 'he was always pestering someone for money.'[14] By 1674 he owed some four thousand pounds, well over £600,000 in terms of today's purchasing power. Perhaps something could have been retrieved by retrenchment and judicious land sales, but Lloyd attempted to restore his fortunes by embarking on a number of costly ventures including mining coal at Llanforda and lead and silver at Minera, near Wrexham. The Cardigan Bay herring fishing business was perhaps his only profitable scheme.[15] After first considering Nefyn, Lloyd realised that he could exploit his family connections with the Pryses of Gogerddan and the family's cadet branches along the river Dyfi. The original plan of about 1674 was to set up a syndicate. A draft prospectus was set out as a questionnaire, a form which natural historians were beginning to use. This was the work of the Shropshire surveyor and cartographer, John Adams (*c*.1670–1738), who subsequently made it the basis of his *Index Villaris* (1680), a catalogue of cities, towns and villages in Britain:

> about eight years since I was at Mr. *Lloyd's* of *Llanvorda* ... who then Designing a Fishery on the Coast of *WALES*, I endeavor'd to compute what *Sale* he might probably make in the *Neighboring Markets* ... making *Aberdovey* ... the first *Landing-place*, I set down all the *Markets* within an hundred Miles.[16]

In the event, the fishery became a partnership between Lloyd and Sir Thomas Pryse of Gogerddan. Much of Lloyd's correspondence in

the 1670s relates to the venture and the acrimonious disputes it generated; a note in a later hand in one of his letter books reads:

> The said Edwd. Lloyd was a very debauched, litigious, quarrel-
> some Welshman; who had been a great sufferer for the Royal
> Cause, but in other respects a most abandoned libertine ... His
> letters ... concerning disputes and squabbles with his relatives &
> others ... show a most peevish, quarrelsome, revengeful and mali-
> cious disposition.[17]

An incomplete and undated letter provides a paranoid (and uninten-
tionally comic) account of the numerous slights Lloyd claimed he had
experienced, first in Aberdyfi and then in Borth and Aberystwyth, in
1675–6:

> Sr Thomas Pryse was very ready in plentifull Cups ... to show me
> a Cordial well come ... & ptended to further any designe I had ...
> But as soone as hee found ye fish were caught in any quantities,
> there did awaken in him a naturall Covetous humor, wch was
> hourly by his Lady & a certain sort of vermine yt crawle about
> him ... stirred wh filled him wth suggestions & bare surmises of
> profit, at wch he greedily grasped ... I began to find yt ye liberty
> Sr Thomas had given me to fish was only a Leave to fish for him.[18]

Lloyd may also have traded in fresh-water fish since archaeological
investigations have revealed the remains of some twenty fishponds at
Llanforda, where three or four would have met the needs of a house
of that size.[19]

Lloyd did not forgive those who had served as tax collectors and
sequestrators during the interregnum. He kept a close eye on Oswestry,
drawing up in 1660 a lengthy and minutely detailed list of 'Oswestry
Malignants' and in 1674 'Articles against ye Phanaticall Corporation
of Oswestry'.[20] Valuable evidence of the strength of Puritanism in
Oswestry, they are even more significant as indicating Lloyd's obses-
sive and unbalanced personality. As party strife intensified in the later

1670s, Lloyd emerged as an eager participant in local elections, an activity which he could ill afford. He was so desperate to support his distant relation Edward Kynaston (1641–93), the unsuccessful Court candidate in the March 1677 Shrewsbury by-election, that he commissioned a list of Lloyds who had been burgesses of Shrewsbury since the reign of Henry VI in the vain hope of establishing his right to vote there.[21] In the February 1681 Denbigh Boroughs election Lloyd, directed by Sir John Trevor (*c.*1637–1717), acted as agent in the attempt to unseat Sir John Salusbury (*c.*1640–84) of Lleweni.[22] In his electioneering Lloyd made use of his legal knowledge, since he had spent considerable time in London during the 1670s reading for the Bar. Although his name does not appear in the registers of any of the four Inns of Court, he wrote to his housekeeper 29 November 1677 that he had been called to the Bar and 'sworn utter Barrister'.[23] With typical Lloydian optimism, he claimed that he would make so much money as a lawyer during the next three years that he would not see the want of the estate he had lost to William Williams (1634–1700), an ambitious and ruthless lawyer who was to become Speaker of the House of Commons.[24] Williams gradually acquired land and mineral rights as Lloyd became a forced seller and, as the latter's finances deteriorated, Williams advanced him money which he knew Lloyd could never repay. Lloyd appealed to friends to save him – 'suffer not me to be a prey to this prodigous glutton who will not buy my land unless for sauce my discretion is served up to his taste' – but by early 1676 he was finally obliged to write that he had flung himself 'into ye Leviathan of Glascoed's mouth', selling the estate to Williams at sixteen years' purchase. Lloyd retained a life-interest in the house, residing there until his death.[25] As his financial difficulties deepened he attempted in 1680 to let Llanforda to Lady Herbert of Powis Castle[26] for his lifetime, providing her with a list of the gardens which indicates their scale and variety: 'ye Tulip Garden, Pigeonhouse Garden, Parlor Garden, Wildernesse, Lower Garden and well, ye command of my fishponds, and her herself a key to ye Physick Gardens and ye fruit Garden beyond ye Nursery'.[27]

Apart from the pleasures of the flesh – Bridget Pryse was only one of numerous mistresses and 'comforters' – Edward Lloyd's main delight was his garden. Like his father, he was prepared to spend extravagantly on it. He showed an informed interest in his plants, seeds, plans and experiments. When he complained to Edward Fuller (*fl.* 1679–*c.*1720) of Strand Bridge, the leading London seedsman,[28] about the quality of the seeds he had supplied, Lloyd claimed that after gardening for twenty-six years he possessed a better stock and could sell Fuller seeds. His correspondence shows that he also dealt with another prominent London seedsman, Francis Dryhurst, and that he was growing fashionable plants, such as melon, asparagus and artichokes.[29] Whether Lloyd was engaged in market gardening for profit or selling surplus produce from his kitchen garden remains unclear. Produce from private gardens certainly found its way to the market, sometimes on quite a large scale.[30] However, as Roderick Floud has recently emphasised, the enormous costs incurred by the kitchen gardens of the upper classes, which were intended to provide their owners with fresh produce for luxury consumption throughout the year, made regular commercial sales of their produce uneconomical.[31]

While he was in London in 1673-4 he kept a close eye on his agent's care of the gardens, and during another stay there in 1677–8 he sent his agent precise instructions for handling tulip bulbs he had sent from there: 'open ye lesser box, and in some cunning place where none but you can come aire ye Tulip roots & putt ym up againe.'[32] Other letters show Lloyd's interest in the productivity of his fruit and vegetable gardens and in redesigning his pleasure garden. He had hoped that 'ye wilderness might have been finished ere this' and sent his motto to be written in gold on the flower pot: 'Dyn a feddwl, Duw a ran; Man purposes, God Disposes; Homo proponis, Deus disponit.'[33] Thanks to his agent's care and the labours of Samson, his African gardener,[34] he could write in 1680 that 'no sight as ever pleases me so well as ye snow drops now in ye Wilderness.'[35] After Lloyd's time all this was swept away as the mansion was rebuilt, the final house on the site being demolished in 1949, and the gardens, on which so much borrowed money had been expended, were drastically remodelled.[36] All that remains today are traces of the fishponds and, perhaps, the descendants of the snowdrops enjoyed by Lloyd and his father.[37]

When he offered Lady Herbert the use of Llanforda in 1680, Lloyd said he was resolved to 'mind nothing but his Books and Garden'. There is no catalogue of his lost library, but a few scientific books can be traced from his correspondence and from references in Edward Lhwyd's earliest botanical notes. The most tantalising evidence of the library's contents is in a letter dated January 1676/7 to Katherine Owen of Porkington, wife of William Owen (1624–1677/8): 'I have sent you Shakespeare, which I shall make bold to send for again, when ye booke binder comes … I shall send you a Catalogue of most of ye plays I have and wt you like shall be at yr service. Shirleys are here and there and not together.'[38]

During the seventeenth century an increasing number of the Welsh gentry became interested in science. Meredith Lloyd (c.1620–1701?) was considered a chemist, and in 1664 was invited by Sir Thomas Myddelton to Chirk Castle, where a laboratory was set up for him to assay local mineral ores.[39] Owen Wynne of Gwydir Castle (1592–1660) assembled an extensive library of books dealing with alchemy and metallurgy.[40] Thomas Glynne of Glynllifon (d. 1648) was an enthusiastic botanist and friend of the London apothecary and botanist Thomas Johnson (1595/1600–1644), and contributed two Welsh plants, *Geum rivale* and *Otanthus maritimus*, to Johnson's revised edition of Gerard's *Herball* (1633).[41]

Contemporaries would have considered Lloyd a 'virtuoso'.[42] This ambiguous term was applied to serious natural philosophers, but also to gentlemen who took an informed (but sometimes shallow) interest in a wide range of fields. Replying to a query in December 1680 about a light recently seen in the sky, Lloyd argued that it could not be a comet; it was in fact the great comet of November 1680 to March 1681 (C/1680 V1), bright enough to be visible during the day. In fairness to Lloyd, the comet puzzled contemporary observers, even Newton being uncertain, when it first reappeared after perihelion, whether there were one or two comets.[43] Writing to the astrologer Dr John Gadbury (1627–1704) to request his opinion about 'the streaming star', Lloyd stressed that his enquiry was 'not of its portend but of its nature', a remark which reflects Gadbury's futile attempt to develop a reformed astrology which emphasised experiment rather than predic-tion.[44] Lloyd's correspondence with Gadbury also covered botanical and political matters (both were high Tories) and enabled him to order new books such as Nehemiah Grew's catalogue of the Royal Society's collection, *Musæum Regalis Societatis* (1681). Another correspondent was Edward Tyson (1651–1708), the founder of modern comparative anatomy, who was the first to establish that the porpoise is a mam-mal.[45] Lloyd called Tyson his teacher in anatomy and hoped he would provide him with a microscope, an instrument in vogue following the publication of Hooke's *Micrographia* (1655).[46] Lloyd also apparently believed in 1680 that he was qualified to set up 'a small shop of phar-macy'. In 1680–1 'Dr Claudius', the 'Salopian Botanick', spent time at Llanforda; as well as selling Lloyd plants and seeds, he carried out expensive experiments there.[47] Dr Frederick Clodius (1629–1702) was born in Schleswig and moved to London in 1651–2 as an agent of Duke Friedrich III of Holstein-Schaumberg.[48] A well-connected scholar, he associated with Commonwealth promoters of intellectual, agricul-tural and educational reform, notably Samuel Hartlib (*c*.1600–62).[49] Following the Restoration his fortunes declined, but in 1664 he was temporarily rescued by William Brereton (1631–80), a mathematician and founding member of the Royal Society. Cast adrift as a result of Brereton's financial difficulties, Clodius settled in Shrewsbury, restoring his fortunes by practising as a physician and acting as a botanical and medical consultant to families such as the Cholmondeleys of Cheshire. Comments and notes in one of Lloyd's notebooks record some of the investigations Clodius carried out at Llanforda in 1680–1:

> Take rain water and boyle to ye Consumption of ye half … to make spirit impregnated with odour of individual flowers.
>
> Claudius.
>
> Things (chemicals and salts) put into ye chest in ye great Room July 7th 81. All these were made or refined by Dr. Clodius at Llanvorda anno 1681.
>
> Unguents and Balsams to be made this Year.[50]

Other important correspondents were the Oxford botanists Jacob Bobart (1596–1680),[51] Keeper of the Oxford Physic Garden, and Robert Morison (1620–83),[52] the Professor of Botany, with whom he exchanged seeds and catalogues. Lloyd also asked Morison for a copy of his latest book, probably *Plantarum historiae universalis Oxoniensis pars secunda* (1680). Perhaps the botanist who had the greatest influence on Lhwyd's early work was Edward Morgan (*c*.1619–88?). Morgan has never received due recognition since, as John Harvey pointed out, he was a practical gardener and botanist who never ventured into print.[53] Morgan's place of birth remains unknown, but the Ashmolean Book of Benefactors notes him as '*Glamorganensis*'.[54] He received plants from John Tradescant (*c*.1570–1638) during the 1630s and was a witness to his will in 1638.[55] A Welsh speaker, Morgan acted as interpreter when members of the travelling club of botanists (the *Socii Itinerantes*), headed by Thomas Johnson, made the first scientific ascent of Snowdon in August 1639.[56] The party called en route at Bodysgallen, where Robert Wynn I was enlarging and remodelling his house and gardens.[57] This may have been Morgan's first visit to the place where he was to end his days half a century later. There they also met Thomas Glynne of Glynllifon. In his account of the expedition, Johnson described Morgan as '*rei herbariæ studiosus*'.[58] Gardener at a physick garden to the south of Westminster Abbey (which may have had some connection with the Society of Apothecaries), Morgan was for several decades a very well-known figure in English botanical circles because of his practical skills and great knowledge of plants.[59] John Evelyn, who visited him at the Westminster Physic Garden in June 1658, described him as 'a very skillful Botanist'.[60] A frequent visitor was John Ray (1627–1705), whom Morgan supplied with seeds and plants for the gardens of Trinity College, Cambridge, in 1662 and 1669, and Morison pronounced Morgan's garden 'the best collection of plants in England'. In 1673 the Society of Apothecaries leased the site of the present-day Chelsea Physic Garden, and the Westminster plants were ordered to be removed there. Morgan was not employed at Chelsea, possibly because he was considered to be too old and expensive.[61] He was still at Westminster in 1677, when the garden was visited four times by the Quaker botanist

Thomas Lawson (1630–91), who listed some five hundred of its plants.[62] Subsequently, with legal assistance provided by Edward Lloyd, Morgan leased the garden to Robert Rusholme and moved to Wales. By 1680 Morgan, though based in Bodysgallen where Robert Wynn II (b. 1655) was extending the house and gardens, was also employed at Llanforda, where he set about comparing the growth rate of plants in Shropshire and London. It was at Morgan's suggestion that Lloyd began buying seeds from Fuller, and he also arranged for him to be sent seeds from the Oxford Physic Garden and from St Christophers and Virginia, activities noted in Lhwyd's hand.[63] Perhaps the most innovative work Morgan undertook at Llanforda was establishing systematic beds, growing plants in thirty-nine sections arranged according to 'Bp Wilkins sections', the classification scheme Ray had provided for *An Essay Towards a Real Character* (1688) by John Wilkins (1614–72).[64] Some of the plants were recorded by Lloyd, others by Lhwyd.[65]

From 1672 onwards Morgan had been compiling a three-volume *hortus siccus* which eventually contained about two thousand plants and flowers, neatly labelled and with an index to the whole collection, all in his own hand.[66] A letter addressed to Edward Lloyd 31 July 1680[67] bears on the reverse, in Lhwyd's hand, a list of plants 'found in north wales mont: nostras J.B.', with cross-references to the appropriate entries in the *hortus siccus*, proof that the volumes were then at Llanforda. Morgan had taught pupils at Westminster and may have brought some with him to Llanforda. He told Bobart that he had sent 'My pupills abroad who already have arriv'd at a tollerable Competency of knowledge in Plants',[68] and in a letter to Robert Morison, 3 December 1680, discussing an exchange of seeds, he also mentioned 'Two pupils yt have last year taken a gt deal of pains in Breddin hills, snowden fforest, severall mountains rocks Cliffs by ye sea side, sea shore in many places, wee searched for ye solidago sarasenica but wee were ill directed.'[69] Lhwyd may have joined Morgan's pupils on their forays, but by 1681 he was botanising independently, his father noting that 'Neddy gone a-simpling' on the estate.[70] A notebook contains Lhwyd's 'Diary of plants (and where found)', with references to finds at local sites such as Llwyn-y-maen and Craig Forda, and from further afield such as the Breiddin Hills, which he subsequently revisited.[71] Later entries in the notebook (pp. 57v–63) are of the greatest importance in enabling us to trace the development of Lhwyd's botanical skills and knowledge and have been discussed in detail by A. O. Chater, who also provides the modern binomial names for most of Lhwyd's finds.[72] The lists

> show sometimes how Llwyd tried to work out the identity of a plant, worrying whether what he had found was the plant that Gerard or Parkinson knew by a particular name, and changing his mind as he realised more satisfactorily what it was … The entries show how interested he was in the ecology of many species, and how particular he was in describing localities. They show what books he seemed to be learning his plants from. And they show how good he was at finding … very rare plants. The impression is that he found the plants himself, rather than being shown them.[73]

The entries cover the period 1681–2. Regrettably, Lhwyd provides the month but not the year for some of them. Thus for 3 June [1681?] he lists over twenty 'plants found upon Snowdon hils in Caernarvan-shire', his first visit there. He places his finds in a broader context, noting that *Saxifraga hypnoides* is found 'Also on Breiddin hil in Montgomery shire. and all ye high hils in North Wales'. He also provides specific detail about where some plants were found, *Oxyria digyna* in 'drie water courses near Llanberrys', *Diphasiastrum alpinum* 'on ye high top of ye Gluder'. Another entry notes four plants found 'In ye ditches before you passe over Severn goeing to Breiddin hill' and another two plants found 'On Gwern Velk', an unknown but clearly wet locality, as indicated by the Welsh place-name element '*gwern*', alder tree, and, by extension, the boggy places where such trees grow, and also by the fact that one of the plants, *Mentha pulegium* or Pennyroyal, flourishes in damp or boggy ground.[74] At this stage he does not provide the Welsh names of his finds.

Since the next series of entries are dated they enable us to trace Lhwyd's travels. On 17 April 1682 he noted eight plants 'found at Aran Benllin about 5 miles distant from Bala in Merioniddshire', including

Saxifraga hypnoides, 'On ye moist places of the Rocks and by ye currents of water which descended from ye mountains, very common', and *Vaccinium oxycoccos* 'on ye boggie grounds about Llaniwllyn'. An additional note on this plant shows Lhwyd seeking information from local people, 'They told me that they had of these berrys that were white as well as spotted and red.' *Meum athamanticum* is given a very precise location, in 'a field close by Llaniwllin called Bryn y ffaenigl' (Hill of the Fennel). He next referred to finds from 'a mountain called Cefn Lwyd on the back side of Craig yr aderyn in Meirionith shire' and from the ruined walls of nearby 'Castell Llanihangel' (Castell y Bere). Five plants are recorded from 'Craig Verwin, in Denbighshire', his first visit to the Berwyn Hills. By 20 May he had reached Cadair Idris, where he noted eleven plants, seven of them in 'ye Rocks about Llyn y Cau where they told me that not many years agoe, an Eagle did Breed', in Chater's view as plausible a record as any of eagles nesting in Wales, and also a further example of Lhwyd recording information from local inhabitants; the other plants he found 'Towards ye top of ye hill'. He subsequently marked the Cadair Idris plants he had also found on Pumlumon. Following this comes a list of 'Plants found in the marshes below Tyno hîr', the house of his mother's brother, Walter Pryse. Lhwyd then crossed to Aberdyfi, noting on 21 May 'Sea Plants observed about Aber Dyfi', some of which he may have found at Ynys-las as he crossed back to Borth and Aberystwyth. Finally, on 24 August 1682 Lhwyd drew up 'An exact account of what plants were then found on Snowdon-hils',[75] a list of twenty-seven plants virtually corresponding to the list in NLW, Peniarth MS 427 (G, pp. 67–9), which he later compiled for the use of David Lloyd (*c*.1660–1703) of Blaenyddôl, Corwen.[76]

Towards the end of his life Morgan ceased to reply to Lhwyd's letters. Writing to David Lloyd in June 1685, Lhwyd wondered whether Morgan 'be dead or alive', adding that he had told him several times he would leave him 'his Studie of Books, wth [worth?] abt £10' (G, p. 74). Lloyd was accordingly charged with assuring Morgan, when he visited Bodysgallen, 'that he cannot bestow em on one that wishes him better, nor perhaps on any Friend yt will make better use of ym'. A year later, following a visit to Bodysgallen in August 1686, John Wynne

(1665–1743) of Maes-y-Coed[77] told Lhwyd: 'we found old Morgan busy in his garden who made us all the wellcome he could, and presented us ye finest flowers in his paradise. He was much troubled that you had disappointed him in not comeing to ye countrey this summer: He seems to be very old and decrepit, and not able to last long.'[78] After Morgan's death[79] his *hortus siccus* came to the Ashmolean Museum and his name was duly entered in the Book of Benefactors.[80]

Edward Lloyd was not an easy man to live with – arrogant, aggressive, rude and outspoken, a chronic invalid at forty-six[81] and prone to bouts of paranoia. When he wrote to Bridget he never referred to the boy as 'our son', nor even as 'my son', but only as 'your son'; she, on the other hand, refers to 'our little boy' and 'my Nedy'. Lloyd appears never to have been well disposed towards his son, writing to Bridget on 1 June 1681:

> I neither have nor had obedience from him wch is so quenched in him yt I find nothing but arrogance and an undervalueing of all people but himself so yt I whether I live or dye must refer to him to Cosen Prise [Thomas Price of Llanfyllin (1627–1704)], who must manage him or he is utterly lost.[82]

Within a few weeks, on 26 July, Lloyd died and was buried on 30 July 'about nyne o'th Clock at night by Torch-light without any Sermon' in the chancel of St Oswald's, Oswestry.[83] In his will, drawn up 23 July 1681 and proved in November, he made fourteen bequests. First-named was his 'wellbeloved friend and kinsman Mr Thomas Price', who was left some of Lloyd's most treasured possessions, such as his silver sword and best case of pistols, together with 'what books and papers hee shall think fit'. The second legatee was 'Mrs Bridget Price', bequeathed ten pounds 'to buy her a Ring'.[84] Edward Morgan was left 'towards what money I owe him and his wages ffiftie pounds, with a Sute of mourning', and John Gadbury received forty shillings to buy a ring. The only specific bequest to 'Edward Sonn of Mrs. Bridget Pryse' was of Lloyd's 'fine lynnen'. That he was not given a suit of mourning or ring may indicate the strained relationship between father and son, as does the wording of the final section of the will: 'I doe alsoe committ the said Edward Sonn of Mrs. Bridget Pryce into the Tuition and care of my said Executors before-named to be dealt withall according to their discretion and his demeanour.' The two executors were Thomas Francis, Lloyd's agent and, as 'Honest Tom', recipient of his letters concerning estate matters, and Thomas Price.[85]

3

OXFORD

In late October 1682, nearly fifteen months after his father's death, 'Edward Lloyd' went up to Jesus College, Oxford.[1] Before this he had probably attended Oswestry School. The school's seventeenth-century registers have not survived,[2] and the tradition that he had been a pupil there apparently originated in Richard Ellis's biography: '[he] probably entered Oswestry Grammar School, and it was not unlikely that he taught in it' (G, p. 2). While not enjoying the prestige of Shrewsbury School, Oswestry provided an excellent education for boys from a modest background such as Lhwyd's slightly older contemporary, Thomas Bray (1656–1730), founder of the SPG and the SPCK and promoter of parochial libraries. Given Edward Lloyd's financial difficulties, sending his son to Oswestry School would have had two advantages; it was local, hence no boarding charges, and the fees were low. One may doubt, however, whether Lloyd would have permitted even an illegitimate son to undertake the menial duties of an usher.

Where Lhwyd had lived during the months following his father's death is unknown, though his plant-hunting records for 1682 suggest he spent some time in Ceredigion with his Pryse kinsfolk. He may also have stayed for part of the time at Llanfyllin with his father's executor, Thomas Price; this would explain Lhwyd's detailed knowledge of the natural history and antiquities of Llanymynech, some nine miles away. Since Price was one of a circle of scholars in north-east Wales and the border who studied and copied manuscripts dealing with Welsh antiquities, Lhwyd's father may have hoped that Price would transmit these antiquarian interests to his son. It was possibly then that Price gave Lhwyd a copy of Geoffrey of Monmouth's 'British History' to transcribe (G, p. 243). Price may have encouraged his ward to go to Oxford and remained in contact with him, though the relationship changed in its nature as Lhwyd developed his own scholarly interests. He eventually developed a low opinion of Price, describing him in 1694 as '*a Gentleman (that we \may/ not \envy/ our worst Friends all the good character they seem to deserve*) of some *learning and Ingenuity* ... but nothing of Candour, and no great share of Iudgment' (G, p. 243).

'Edward Lloyd' was entered in the Jesus College buttery books for the week beginning 27 October 1682 and matriculated 17 November. Undergraduates were 'commoners', usually the sons of gentry and clergy in comfortable circumstances, or 'servitors', who were hired by other students or Fellows to wait upon them for their fees and maintenance; between these came the 'battelers', the sons of less affluent gentry and clergy who received their 'commons', but paid for additional provisions, the 'battels'. Lhwyd belonged to this category. Undergraduates followed a four-year course leading to a BA degree, but candidates for a bachelor's degree in a higher faculty such as law or medicine were required to attend lectures for five years after two years' study of *literae humaniores*, the initial BA course. Law students of more than four years' study (the equivalent length of the BA degree) were entitled to be recognised as SCL (Student of Civil Law) and enjoyed the same status in the university as Bachelors of Arts, though remaining undergraduates. The Jesus College buttery books note Lhwyd as SCL in 1686, four years after his matriculation. This would have given him some official status in the University. It did not indicate that Lhwyd intended to pursue his legal studies, since he was by then employed by the Ashmolean.

When he went up to Oxford, Lhwyd's prospects were bleak. He had no home, could expect no inheritance and had no patrons to advance his career. He seems to have received a little money from what Price could salvage from his father's ruin, for on 30 January 1684 Price told him:

Your fathers debts grow so fast vpon me that like Hidra's heads I no sooner satisfie one but two or three new ones start up in the place of it; wch has made me so bare of money that I had much adoe to get this £7 (wch I have sent you) … I have no more but to wish you good successe in your studies.

Lhwyd would have to rely on his own talent and industry to make his way in the world.

No record survives of Lhwyd's student life. He was in regular residence apart from a few weeks in 1683, 1686 and 1687, but his name does not appear in the buttery books after 1687, and he never completed his degree. From early in his university career he found frequenting Oxford's Physic Garden far more congenial than studying law. The Physic Garden derived initially from a desire to nurture and study plants useful for medicinal purposes. The gift of Henry Danvers (1573–1644), first Earl of Danby, who may have been inspired by gardens seen while in exile abroad, it was officially founded 25 July 1621.[3] Since the five acres of ground opposite Magdalen College required much work to raise its level above the Cherwell and to construct high walls around the site, it was not until the end of the 1630s that planting could commence. Danby had wished to appoint John Tradescant, senior, as superintendent, but he was terminally ill, and sometime before 1641 Jacob Bobart, senior, had been offered a lease on the site. He would receive £40 a year as *Horti Praefectus* and would also enjoy the income deriving from sales of the garden's produce. As Danby's estates were sequestrated during the Civil War, Bobart had to manage without his salary for several years by selling produce raised in the Garden. Nevertheless, he succeeded in planting the garden with rare plants and prepared its first catalogue, *Catalogus Plantarum Horti Medici Oxoniensis*, published anonymously in 1648, which listed 1369 plants. A second catalogue (1658), which included five hundred additional plants, indicates how quickly the Garden was being developed. Its status was enhanced when Robert Morison was appointed Professor of Botany in 1669 with a salary of £40 a year. Morison set about reorganising the layout of the garden by introducing the partitioned plots he had seen and used in France.[4] This was the Garden, illustrated in David Loggan's engraving of 1675, that Lhwyd knew.[5] From September 1670 onwards Morison lectured in the Garden three times a week during five-week periods in spring and autumn. By discussing specimens laid out on tables before him, much as anatomists combined description and dissection, he made the Garden 'a teaching resource that might rival the gardens of Leiden or Montpellier'.[6]

Following his father's death in 1680, Jacob Bobart, junior (1641–1719), worked harmoniously with Morison. Although he did not succeed Morison as professor when the latter died following a street accident in November 1683, Bobart continued Morison's research and teaching.[7] He also developed the herbarium commenced by his father, now a founding collection of the Oxford University Herbaria.[8] Although essentially a practical gardener, Bobart was considered to be enough of a scholar to be paid four hundred pounds (well over £60,000 in today's purchasing power) to complete the third volume of Morison's herbal.[9] In 1710 Conrad von Uffenbach (who mistakenly believed Bobart to be a professor) was startled by his poor clothing and worse hat, his 'hideous features and generally villainous appearance', but recognised him as a 'good and honest man', praising his industry 'in publishing the works of his predecessor Morison, who far excelled him in learning'.[10]

It is not known whether Lhwyd had met Morison before going up to Oxford. Although Edward Lloyd had spoken of his intention to take 'Neddy and Mr Morgan' with him to Oxford in February 1680,[11] there is no record of such a visit. Lloyd's correspondence with Morison and Bobart, discussed in Chapter 2, would have made Lhwyd known to them. He would have been welcome as an experienced botanist familiar with the literature and systematic beds, though Morison dismissed Ray's Tables, the scheme Lloyd had employed at Llanforda, as 'a chaotic muddle of plants'.[12] Morison's untimely death limited his influence on Lhwyd, but the latter and Bobart, junior, soon established an excellent working relationship which became a lifelong friendship. Bobart particularly valued Lhwyd's plant-hunting contacts in Wales, and Lhwyd was soon obtaining specimens from north Wales for the Garden, identifying himself with the Garden to the extent of using 'we' and 'us' when writing about it (G, pp. 74, 75).

Lhwyd's correspondence with David Lloyd illustrates the process of acquiring plants for the Garden. Unfortunately the most important letters are undated, but the general picture is clear: Lloyd was regularly supplying the Garden with plants, while Lhwyd sent him rare plants and seeds in exchange:

> I have sent you some small requital of your kindenesses; being a parcel of young trees & shrubs, some very choice ... wth a few flower seeds ... The Virginia Cedar is a plant recently come from yt Country; & I am confident was never in Wales before. I must \desire/ your usual trouble of furnishing us with your Mountain Plants. (G, p. 75)

Lhwyd gave Lloyd precise instructions which plants were required and where they could be found. He added more detailed site information to his 1681/2 notes (discussed in the previous chapter) in the 'Directions ... on Travailing our Hills' he drew up for Lloyd some time later in the 1680s.[13] For Aran Benllyn, the general location 'by ye currents of water which descended from ye mountains' was replaced by the more specific 'By ye rivulets that run through ye rocks above Llyn Llymbran'. Similarly, the mention of an eagle nesting above Llyn Cau was replaced by a specific location, 'on a rock they call Craig dhû, above Llyn y Càu'. To the Snowdon notes he added, 'I observed several other plants, wch because they were not then in flower, I knew not whither to reduce', a problem he was frequently to encounter. He also added a symbol to mark plants 'not observ'd by Mr Ray to be Natives of England'. The 'Directions' may explain why Lhwyd was such a successful plant hunter since he stressed that

> on most of these high hills, ye rarest plants & greatest variety are to be met with, by the rivulets of water that descend from the rocks from ye tops of 'm. In goeing up ... you must make use of a Guide; who must not direct you the easiest way of goeing up; but must bring you to all the steep & craggie cliffs, yt are, (tho but difficultly) accessible. (G, p. 69)

A similar emphasis on tracing rivulets uphill 'as far high as he can with safety' appears in an undated and unaddressed letter in which Lhwyd set out detailed instructions for collecting on Cadair Idris, on 'Rennogl Fawr' (perhaps Rhinog Fawr) and on the coast of Merionethshire. Instructions for collecting were followed by directions for packing plants in damp moss for carriage: 'If these directions be but pretty well observd; we doubt not of receiving as choice a collection as ever our physick Gardener has, from ffrance, Italy or Germany, whence he receivs several boxes of plants every summer' (G, p. 73). Lhwyd's letters also reveal a broadening of his interests from botany to conchology and mineralogy. He told Lloyd:

> Put up all Sorts of Snayl Shells you meet with; all Sorts of River Muscles, or any other Sweet Water Shell.
> Be sure you send some of yr lead oar, & any other fossil yt you can meet with, wch is not every where common; as Chrystals, marcasites, cochlites &c. (G, p. 69)

Lhwyd was using 'fossil' here in its original sense of 'anything dug out of the ground' rather than in the modern sense. In July 1685, after specifying which 'mountain plants' he wished Lloyd to send, he added:

> I have yet an other request to make to you, wch will perhaps at first seem ridiculous … In ye Royal Society at London they have a collection of abt 600 Eggs … to the end, that haveing such a collection before them, they might draw some usefull observations concerning ye shape, size, colour &c. of Eggs in general, for ye improvement of real Knowledge … I desire you'd get some boys to bring you in all \Eggs/ yt y meet with … I would desire but 2 Eggs of a kinde … you must write in yr Letter how ye birds are called in Welch. (G, pp. 75–6, date from *EMLO*)

Once the eggs had been blown they should be packed for transport in 'some pitifull litle basket with hay or fine mosse betwixt each Shell', or perhaps 'a little wool, \feathers/ or plû'r Gweunidd [cotton grass] would doe better.' This letter is of particular interest as an early statement of Lhwyd's approval of the Baconian approach discussed below. His request for Welsh bird names is also notewothy. The Physic Garden was the first University institution to be devoted to the pursuit of science. Closer to Jesus College, Lhwyd could see the University's second scientific premises, which were to provide him with a base for his life's work, being completed on Broad Street, some three hundred yards away. Much of the shell of the Museum had been completed by the autumn of 1680,[14] and when Lhwyd went up in 1682 internal work on the building was well advanced.[15] Histories of the Ashmolean Museum usually begin by outlining the history of the Tradescant collection, its acquisition by Elias Ashmole (1617–92) and his transfer of it to the University. Such accounts are coloured by the Ashmolean's later fame as a museum and art gallery. Contemporaries tended to emphasise the experimental aspect of the institution, Lhwyd writing in 1686 of the 'sumptuous new Buylding wch we have here at Oxford calld the Chymistry' (G, p. 76) and explaining a few years later that items directed to the Museum miscarried since 'the generality of the people at Oxford doe not yet know, what ye Musæum is; for they call ye whole buylding ye Labradary or Knaccatory' (G, p. 133). A less anachronistic approach is to explain why Oxford thought it necessary to set up, at great expense, an institute for scientific research.

As Feingold has shown, rather than being a stronghold of dogmatic Aristotelian scholasticism, seventeenth-century Oxford was prepared 'to honour Aristotle and yet forsake him whenever he was in the wrong'.[16] Unlike other European universities, Oxford never issued any official condemnation of new scientific ideas,[17] displaying an intellectual pluralism based on an awareness that no single contemporary natural philosophy could ultimately explain all observed phenomena.[18] Although new ideas in areas such as astronomy became firmly established during the first part of the century, the rise of experimental science at Oxford dates from the interregnum, the key figure being John Wilkins. Wilkins had been a leading figure in informal London groups discussing science during the 1640s and, following his appointment as Warden of Wadham College in 1648, he organized a club which met in the Warden's Lodgings. Members, some of whom moved to Oxford to participate in its discussions and experiments, included John Wallis (1616–1703), William Petty (1623–87), Robert Boyle (1627–91) and Christopher Wren (1632–1723). Although Wadham College acquired chemical apparatus, the club tended to discuss astronomy, optics, mechanics and medicine. Much of the pioneering work in chemistry, such as the air-pump experiments of Boyle and Robert Hooke (1635–1703), was carried out in private premises, often in the laboratories of Oxford apothecaries and alchemists.[19]

Following the Restoration many natural philosophers returned to London, some participating in founding what was soon to become the Royal Society. Despite their diverse interests, they were in broad agreement with the Baconian precept that the development of natural science demanded specific, repeated observation and controlled experiment, and they asserted the inductive method of proceeding from the specific to a general statement and thence to a hypothesis. The Royal Society's motto, 'Nullius in verba' ([take] the word of no one), expressed a rejection of established authorities and the need to verify all statements by an appeal to facts determined by experiment.[20] They believed that scholarship progressed by sharing ideas expressed in plain language rather than through the secrecy and wilfully obscure terminology of alchemists. Societies were the channels for disseminating this knowledge through lectures, experiments, discussion and, above all, publication. Despite a chronic shortage of money,[21] the Royal Society built up a classified collection of natural objects at Gresham College and published a catalogue of it in 1681 (one of the last books acquired by Edward Lloyd), amassed a library, demonstrated experiments and, from March 1665 onwards, published its *Philosophical Transactions*. In the later 1660s it hoped to construct a 'college' to provide a stronger institutional focus. Wren drew up a plan for a building which anticipated many features of the Oxford Museum, but the scheme was far too ambitious for the Society's limited resources.[22]

Natural philosophers in Oxford shared a similar vision of organised scientific research which, they hoped, would be financed by the University. As Ovenell points out, Ashmole's offer was made when the most influential University figures were 'well disposed to welcome it ...

even without Ashmole's offer such an institution would soon have materialized'.[23] As early as 1653 Wilkins had offered two hundred pounds towards establishing 'a college for experiments et mechanicks', but other donors failed to come forward.[24] In 1664 Boyle was told of a scheme for endowing 'a society to study experimentall philosophy at Oxford', but it was not pursued. A third attempt about 1670 reflected Oxford's resentment of what were considered to be the pretensions of the Royal Society, exacerbated by a bitter and lengthy dispute over manuscripts bequeathed by the Duke of Norfolk to the Society. In August 1670 John Evelyn told a correspondent that at Oxford 'they talke already of founding a Laboratorie, & have begged the Reliques of old Tradescant to furnish a Repositary.'[25] By 1677 the influential Vice-Chancellor Dr John Fell (1625–86) had come to support the project as part of his scheme to enhance the reputation of the University. A further impetus was provided by the publication in that year of Robert Plot's *The Natural History of Oxford-shire*, which prompted discussion of plans for

erecteing a Lecture for Philosophicall History to be read by the author of that booke; to which end, as soon as we are agreed on the ground, we shall build a school on purpose for it with a labratory annext and severall other rooms ... whereof on[e] is to hold John Tredeskin's raritys, which Elias Ashmole ... hath promised to give to the University as soon as we have built a place to receive them.[26]

The master-gardener John Tradescant, senior, made good use of the patronage of his aristocratic and royal employers to amass a remarkable collection of 'curiosities', which was maintained after his death by his son, John Tradescant, junior (1603–62). By 1628 Tradescant, senior, had bought a substantial house and garden in Lambeth, where he housed his collection and Museum of Antiquities.[27] Open to visitors for a small fee, it became a popular attraction. Many educated gentlemen in the seventeenth century assembled such 'cabinets of curiosities'. Their research value varied greatly since their contents could range from random collections of antiquities and exotic objects to materials systematically brought together for scholarly study,[28] but, as private collections, all risked dispersal when their owner died. The death of his only son in 1652 accordingly made it necessary for Tradescant, junior, to decide the future of his collection.

Ashmole had first come into contact with the Tradescants in 1650. He was an able, self-made man who, as a favourite of Charles II, became Comptroller of Excise.[29] A founder member of the Royal Society, his scholarly interests ranged from natural history, mathematics and astronomy to alchemy and astrology (he rarely took an important decision before casting a horoscope), antiquities, numismatics, genealogy, heraldry and the orders of chivalry. As an avid collector of books, manuscripts, coins and curiosities he had the experience to compile the first printed catalogue of Tradescant's collection, *Musæum Tradescantianum* (1656), a second edition appearing in 1660. Probably as an acknowledgement of Ashmole's work, Tradescant arranged in 1659 to transfer the collection to him by a Deed of Gift, but this was subsequently contradicted by the terms of his will. Soon after Tradescant's death in April 1662, Ashmole embarked upon a legal dispute with Hester, Tradescant's widow, and in May 1664, the Court of Chancery ruled in favour of the precedence of the Deed of Gift. Even so, Hester Tradescant was to keep the specimens in trust during her life. When Ashmole came to believe that she had sold a number of exhibits, he leased an adjacent house in the autumn of 1674 to prevent further losses. Following an attempted burglary, Hester reluctantly transferred custody of the collection to Ashmole in the summer of 1675, though it did not pass fully into his ownership until she was found drowned in her garden pond at the beginning of April 1678.

Although Ashmole had never matriculated, in 1645 he had been briefly associated with Brasenose College, Oxford, and had developed a strong attachment to the University. In the summer of 1675, with the Tradescant collection now under his control, he opened negotiations with the University about its transfer and that of his own collection. The discussions were protracted, since his original proposal for building a 'large Roome, which may have Chimnies, to keepe those things aired that will stand in neede of it',[30] became a plan for a 'school' for the study and teaching of natural philosophy with its own laboratory, lecture room, library and collection of specimens.[31]

FIGURE I North front of the Ashmolean Museum

The building, designed and built by the local master-mason Thomas Wood (*c.*1643–95), was a deceptively simple structure, its public areas consisting of three large halls, one above the other.[32] The lowest, below street level, was the chemical laboratory containing up-to-date apparatus and furnaces. Its strong vaulted stone roof protected the upper floors from fire or explosion. The entrance, into the second hall, was through an ornate and impressive portal at the east end facing Wren's recently completed Sheldonian Theatre and the quadrangle bounded by the buildings of the Bodleian Library. An inscription above the entrance proclaimed the building's three functions: *Musaeum Ashmoleanum, Schola Naturalis Historiae, Officina Chymica.* The middle floor was to be the lecture hall, the *schola naturalis historiae*, a title that reflected those of the corresponding *scholae* in the Bodleian Old Schools quadrangle and one that emphasised the university role of the building. The collection of specimens, arranged in cabinets, was on the top floor, where the library also occupied a number of rooms. The two upper floors were linked by a great staircase. The building cost the University over £4,500, over £725,000 in today's purchasing power. The University's coffers were so depleted by its construction that for several years the Bodleian Library claimed it could not buy any books.[33] Despite its imposing appearance the Museum building was notoriously damp; in December 1691 John Aubrey (1626–97) wrote to Lhwyd: 'Mr Ashmole and I doe both desire you, to let the Pictures hang *reclining* from the Walls: otherwise the Salt, and Saltpetre in the Walls will rott the Canvass; as you have a Sad instance of the Queens picture in the roome \within/ by the Laboratory.'[34] Later scientists analysed the efflorescence, and a specimen of 'Nitre from the walls of the Museum', collected in the early nineteenth century, still survives.[35] Contemporaries attributed Lhwyd's premature death to his sleeping in a damp chamber in the Museum.[36]

The collection was conveyed to Oxford in February and March 1683.[37] By early May the exhibits were in place, ready for the grand opening on 21 May 1683, but it was not until 26 May, when Ashmole formally transferred his gift to the University, telling the Vice-Chancellor that from his 'Duty and filial Respect, to my honoured mother the University of Oxford', that both the Tradescant collection and his own were donated to the University. There is some doubt about which specimens from Ashmole's collection were given,[38] since many had been destroyed or damaged by a fire in his chambers in January 1679.[39]

Other donations soon followed, the most notable – second only to Ashmole's benefaction – being those made by Martin Lister (1639–1712).[40] A Cambridge graduate, Lister spent some years pursuing his medical studies in Montpellier before setting up a practice in York in 1670. He soon became recognised as an outstanding natural historian, his interests ranging from spiders, shells, fossils and chemistry to antiquities. Lister's very generous gifts to the Museum from September 1682 onwards, ranging from Roman altars to shells and over 1,250 books, were not altogether altruistic. He hoped they would prompt Oxford to confer upon him an honorary MD, which would license him to set up a practice in London.[41] Lister moved to London in September 1683, in anticipation of the degree, which was conferred upon him the following March. His medical career flourished, culminating in his being appointed physician to Queen Anne in 1709. Once in London Lister rapidly established himself in scientific and antiquarian circles and became friendly with Ashmole, a connection which was to aid Lhwyd's attempt to succeed Plot as Keeper.

During 1683–4 Ashmole drafted 'The Order & Rules for the Ashmolean Museum in the University of Oxford'. The preamble stated clearly that the Museum was intended to express the spirit of the new philosophy:

> Because the knowledge of Nature is very necessarie to humaine life, health and the conveniences thereof, and because that knowledge cannot be soe well and usefully attain'd, except the history of Nature be knowne and considered; and to this, is requisite the inspection of particulars, especially those as are extraordinary in their Fabrick, or usefull in Medicine, or applyed to Manufacture or Trade.[42]

Ashmole was acutely aware of the need for security in collections open to the public and in the definitive version of his Statutes, issued in June 1686, set out strict regulations for administering the new institution, 'Lest there should be any misconstruction of my Intendment, or deteriorating of my donation'. Responsibility for the security and maintenance of the collections was vested in six *ex officio* University Visitors, who would annually examine their condition and development and inspect the Museum's financial accounts. To ensure Visitations took place, the sum of six and a half guineas a year from the profits of the Museum was to be spent on 'entertainment' or gloves as an honorarium to them. The Visitors were the Vice-Chancellor, the Dean of Christ Church, the Principal of Brasenose, the Regius Professor of Medicine, the Senior Proctor and the Junior Proctor. To facilitate their work the collections were to be categorised under six headings, as described in Chapter 4, and a catalogue was to be prepared for each Visitor to serve as an inventory of his category. The director and administrator of the Museum would be the Keeper, assisted by the Underkeeper (*Procustos*) and a general servant. The Keeper was to receive up to fifty pounds a year, the Underkeeper up to fifteen pounds and the servant forty shillings. Since Ashmole did not provide an endowment for these posts, staff salaries and the substantial honorarium for the Visitors had to be met from fees paid by users of the Museum, which usually fell short of what was required.[43] Ashmole clearly regarded the Museum as a monument to himself and his interests, rather than as a University centre for science teaching and research. The University for its part had met the substantial capital outlay for the Museum's buildings and equipment but was not prepared to finance its ongoing costs. From the start the Museum's effectiveness as a research institute was fatally compromised.

The first Keeper, Robert Plot (1640–96), was Ashmole's own choice. Plot had graduated BA at Magdalen Hall in 1661, MA and BCL in 1664, DCL in 1671.[44] He became Vice-Principal and Tutor at Magdalen before moving to University College about 1676. Plot was appointed Keeper and the first University Professor of Chemistry in 1683, a dual appointment which was not continued following his resignation as professor in October 1690. An erudite scholar pursuing the wide range of interests characteristic of followers of the new philosophy, Plot had participated in the developments in chemistry in Wadham College while retaining an interest in alchemy.[45] In the early 1670s he had drawn up plans to tour England to gather data for a comprehensive 'natural history'.[46] Although the plan was far too ambitious, it led to his influential *The Natural History of Oxford-shire* (1677), the first comprehensive account of the local history, antiquities, natural history and folklore of an English county. Unlike the work of earlier chorographers which concentrated on the genealogy and heraldry of county families, this provided a detailed account of the natural history of the region based on fieldwork, his personal observations being supplemented by questionnaires. The Royal Society recognised the quality of his book by electing him a Fellow in 1677. *Oxford-shire* also convinced Ashmole that Plot was a proper person to be entrusted with his Museum, Plot's secret alchemical interests possibly being an additional recommendation. The book's successor, *The Natural History of Stafford-shire*, was published in 1686, and Plot planned a similar book on Middlesex and Kent in the 1690s.

The University believed that Ashmole had agreed to endow the Chemistry chair, but since he did not do so,[47] the running costs and salaries of the Professor and his technician, the 'Operator', had to be met from fees for demonstrating experiments, from teaching and from the sale of medical preparations. Plot lectured three times a week in the *Schola*, where his discourses could be illustrated by experiments demonstrated in the basement below by the Operator, Christopher White (1651–96), and by examination of specimens and samples on the upper floor. Much of the experimental work was related to pharmacological preparations, and the medical link was emphasised by the use of the Museum basement for anatomy teaching.[48] When Plot's successor, Edward Hannes (*c.*1664–1710), left Oxford around 1695 to pursue a lucrative medical career in London, the unendowed Chemistry professorship lapsed.[49]

Plot embarked energetically upon his responsibilities as Keeper. The most immediate need was for the steadily growing collections to be classified and catalogued within two years of the receipt of Ashmole's donation, as the Statutes stipulated. Since this would have been an impossible task for Plot to undertake single-handed, he appointed two undergraduate assistants in 1683 or 1684. One, Obadiah (or Abdias) Higgins, aged twenty-one when he matriculated 14 November 1684, may have been known to Plot as a student at Magdalen Hall. His main duties seem to have been routine housekeeping and supervisory tasks.[50] The other was 'Edward Lloyd'. It is not known when or how he first came into contact with Plot, but the latter recognised him as an intelligent and experienced student of natural history and appointed him as an assistant, on a meagre salary. Lhwyd shared some duties with Higgins, mainly the supervision of visitors and collecting fees, but his main task was to organise and catalogue the collection. Unlike Higgins, who combined his duties at the Museum with his studies, Lhwyd seems to have devoted himself entirely to the Ashmolean. Though listed in the Jesus College buttery books up to 1687, he had come to regard the Museum, rather than Jesus, as his Oxford base, 'Ed. Lloyd. Mus. Ash.' being listed as a Pikeman in the muster of the Company of Foot raised by Lord Norris at the University at the time of the Monmouth Rebellion in 1685.[51]

APPRENTICE YEARS
AT THE ASHMOLEAN

Although Lhwyd later expressed a rather jaundiced view of his years as an assistant at the Ashmolean, they were an invaluable apprenticeship. His cataloguing work introduced him to the problems of identifying and describing accurately a far wider range of objects than he had hitherto encountered. More significantly, his work brought him into contact with leading scholars, in person and by correspondence.

Despite several short-lived attempts in the 1690s to publish learned journals, the only established English scientific periodical was the *Philosophical Transactions* of the Royal Society. Much scientific communication still relied on word of mouth, formally at meetings of bodies such as the Royal Society and informally in settings such as coffee houses. When face-to-face communication was impossible, scholars had to rely on developing a network of correspondents. For a novice scholar, being invited or permitted to open a correspondence with an established scholar was a formal rite signifying acceptance into the scholarly community. For established scholars, such an invitation from a peer represented a significant recognition of their status; as Lhwyd told Lister in March 1696, 'I receivd about a week since a Letter from Rivinus of Lipsick [Augustus Quirinus Rivinus (1652–1723)] who offers a Correspondence; and to communicat any of the Fossils of Germany … I think it best to close with him' (G, p. 321, date from *EMLO*). Maintaining a correspondence could be expensive since the carriage of letters was paid by the recipients. Lhwyd told Aubrey in April 1692, 'As slender as my fortune is, I shall not think it much to pay for your letter; but shall take it as an honour (besides ye benefit I receive thereby) when ever you please to write' (G, p. 162). One correspondent might send excerpts from an interesting letter to another or even forward the original letter; Lhwyd, for example, sent John Lloyd a letter from Ray, asking for its return for preservation 'because I poste up all his Letters' (G, p. 271). Scholars with similar interests might never communicate directly, relying on an intermediary to keep them informed about their respective discoveries. Thus Ray told Lhwyd in 1699, 'by what I find in your Letters ... of Dr. Richardson's discoveries & observations, I perceive him to be an extraordinary person, of great industry & deep insight into the \more recondite/ History of Nature ... some correspondency with whom, were I not now upon ye Pits brink, would be very desirable.'[1] Letters would often be accompanied by reciprocal exchanges of specimens or books or other favours. Thus when Lhwyd suffered a painful bout of illness mid-November 1692, he asked for Lister's advice, 'haveing no such Interest with the Oxford physicians, as to expect their advice gratis' (G, p. 56), and a few years later Lister asked Lhwyd to act as a mentor at Oxford to his rather unsatisfactory son.[2]

In 1683 Plot established the Oxford Philosophical Society, modelled on the Royal Society of which he was then Secretary. The fourteen or so original members were augmented by a further twenty-one elected in 1684, and subsequently by others. Membership was, in theory, confined to graduates. As well as corresponding with the Royal Society and the Philosophical Society of Dublin, it welcomed letters from collectors and scholars such as Martin Lister, Tancred Robinson (1658–1748) and William Cole of Bristol (c.1622–1701). It met weekly to hear and discuss papers on a wide range of topics of interest to natural philosophers – fossils and botanical specimens, chemistry and scientific apparatus, anatomical observations, extreme weather, antiquities and ethnographical curiosities. The Society met regularly during most of the 1680s but had lost impetus by 1690, Plot describing it as dormant in January 1691 (G. p. 133) and as the 'quondam [former] Oxford Society' in 1692 (G, p. 170).

Though it had met from time to time since February 1683, the first formal session of the Society was held in the *Schola* on 26 October 1683, when Plot read 'a learned Discourse on Earths'.[3] It is not clear when Lhwyd first attended its meetings. A reference in the minutes to a presentation by Lhwyd on 9 December 1684 describes him as 'Mr Lloyd, Register to the Chymicall courses of ye Laboratory of Oxford',[4] suggesting that he was formally one of Plot's assistants at the Museum. Since the Museum Statutes made no reference to such a post, Plot may have created the title to justify Lhwyd's presence, much as Jacob Bobart, junior, had attended ex officio as Superintendent of the Physic Garden.[5] In fact, the graduate rule was not strictly enforced, Cole, a non-graduate but a prolific correspondent, being elected in March 1685.[6]

The subject of Lhwyd's presentation was a sheet of incombustible paper made from 'ye Asbestus-Stone', sent to him by a correspondent in Anglesey, probably Robert Humphreys (1646–1707).[7] Lhwyd may have requested the stones, since the Society had recently shown an interest in 'salamander's wool' and, at its request, John Ballard, a Fellow of New College, had experimented with amianthus (chrysotile) from Cyprus.[8] As will be shown in later chapters, Lhwyd was always aware of the potential economic value of his discoveries and may have hoped to draw attention to a source of a valuable mineral. To produce the paper Lhwyd pounded the stones before taking the powder to a paper mill to be 'mixt with water in their troughs', where it ran together like paper pulp but required frequent stirring because the fibres were so heavy.[9] Lhwyd may have been the first to produce asbestos paper, but 'to be fit for any use' the paper needed to be finer, whiter and stronger. The invitation to present a paper must have come from Plot himself and indicates both his regard for Lhwyd and his own interest in the topic, since a week later Plot demonstrated a superior cloth made from asbestos 'flax'.[10] The experiment provided Lhwyd with the material for his first scientific paper published in the 1684 volume of the *Philosophical Transactions*.[11] It may also have been responsible for Lhwyd's lung problems from the late 1690s onwards, discussed in Chapter 10, since he had exposed himself to finely powdered asbestos.[12]

Lhwyd's next communications were more directly related to his work in the Museum. On 8 December 1685 Plot showed 'some little stones found by Mr Lloyd on a bank by ye wayside' south of Islip church,[13] and on 15 December 'Mr Lloyd communicated some stones like ye Lapides Iudiaci, & others like Shell-fish, which were gathered in this County.'[14] On 12 January 1686, Lhwyd 'presented a catalogue of ye Shells in the Musæum Ashmole' (discussed below), the minute book noting (in Lhwyd's hand) that it 'contains about 700 species of shells'.[15] On 9 March he communicated a sinistral snail shell, found in Cumnor Woods.[16] A few weeks earlier he had communicated 'a paper containing an account of some plants, which grow in North Wales, & are omitted in Mr Ray's catalogue',[17] the second edition of Ray's *Catalogus Plantarum Angliae* (1677), which included plants Ray had found on Cadair Idris and Snowdon during his expeditions to Wales in 1658 and 1662.[18] Henceforth material from Wales came to the fore as Lhwyd reported on his own finds and those of his Welsh correspondents. On 22 February 1687 he communicated 'Curiosities ... sent out of ye Isle of Anglesey' by Robert Humphreys. As well as 'Sea plants & shells', these included specimens of ray egg cases, '*Cîst y Môr i.e. Costa marina*' ('sea chest'), dogfish egg cases, 'their corner strings are not playn as in the former, but curled', and 'Husks of a sort of Sea Insects call'd in Wales *Chwaun y Mor i.e. Pulices marini*' ('sea fleas').[19] Perhaps the most interesting item was a 'Favus marinus', a 'sea bean', or seed carried across the Atlantic from the West Indies to the western coasts of Britain.[20] Robert Sibbald (1641–1722) had recorded in his *Scotia Illustrata* (1684) that these were found on the coasts of Scotland;[21] Lhwyd, demonstrating that he kept abreast of recent publications, produced an example from Anglesey. Lhwyd's 'kind acceptance' of these items encouraged Humphreys to send him a letter 'On Natural Curiosities observed in Anglesey', communicated to the Society in June 1687.[22] Here Humphreys observed, 'I am induced to suspect yt this Island may afford the best entertainment for a Naturalist of all places of its extent & situation in our Country', and hoped that Llwyd would soon visit Anglesey.

On 27 March 1688 Lhwyd communicated 'Some Curiosities' including 'several Roman coyns found at Craig Lhan y Mynych Denbigshire, Some curious Pearles from the River Teivi Cardiganshire, Large Chrystalls from Creigiau'r Eryn [Eryri] in Carnarvonshire ... with both ends entire, Very small Chrystalls from Cardiganshire of the same figure ... but much clearer'.[23] It is not known whether Lhwyd himself had found any of these, but he had visited Snowdonia in 1686–7 and was familiar with Llanymynech, where finds of Roman coins have been recorded over the past three centuries, most recently in 1965.[24]

An early priority was expanding the Museum collections. As Lhwyd told David Lloyd, the Museum '\being/ but lately founded; we are collecting all natural things we can, from all parts to furnish it' (G, p. 76). Over the next few years Lhwyd encouraged friends in Wales to send him specimens. Erasmus Lewes (1663–1745) gathered sea plants and shells 'with spade & basket' on the Cardiganshire coast in July 1685. Although he followed Lhwyd's instructions to pack them in a box with layers of wet moss, the delay while waiting for the carrier rendered them 'very much out of order, and consequently uncapable of doing you any service'.[25] Several of Lhwyd's contemporaries were very ready to contribute to his researches, partly from shared scholarly interests but even more from friendship. His correspondence with them often contained practical advice, but sometimes expressed a broader vision. In January 1686 he sent David Lloyd 'a small collection of shells and form'd stones' as a token of gratitude for the 'many favours I receiv'd from you' and to encourage him in his studies (G, pp. 77–80). The letter then became an apologia for the study of natural history and an eloquent statement of Lhwyd's attachment to the subject:

> I \know/ these & all such like things are generally look'd upon as trivial & unworthy our considerations, but if we consider upon what motives they are \thus/ undervalued we shall finde but small reasons to be discourag'd from our inclinations.

'Critics' might aver that 'to know ye grasses of ye feild; ye Common Stones and Snayls' was 'mean & simple'; secondly, that 'such studies bring us noe profit'; and thirdly, that '\not/ one man amoungst ten hundred men of Learning ... heed any thing of this nature.' Replying to the first objection, Lhwyd maintained that 'ye common Plants, Stones, Shells &c are scarce lesse valuable in themselves; than wheat or rie, rich gemms, and pearls; since 'tis not ye intrinsic worth of things, but ye use men put them to, that makes [the]m valuable'. As for the second objection, 'Gentlemen & others who have sufficient estates, may if they please make these their main Studies since 'tis noe point either of Religion, Moralitie, or humane reason to propose ye getting of money to be ye end of all our Endeavours.' He concluded,

> As to ye 3d obj. that seems to deserve our attention least of any, ffor if men had been allways content to know onely such things as were allready discover'd to their hands, learning could have made noe progresse, & ye world must have been as blinde now as it was two thousand years since.
>
> Soe yt all things examin'd, we shall finde noe reason why men should carpe at these sort of Enquiries, unlesse it be that common error ... of enveying against & condemning most such things as they understand least.

He then described 'the infinite pleasure' that the study of Nature affords, concluding that 'it dayly manifests ye incomprehensible power of our Creator.' The letter closed with Lhwyd's early views about the origins of 'formed stones', outlined in Chapter 6, that they were *lapides sui generis*, stones spontaneously formed by nature.

David Lloyd's younger brother, John (1662–1726), would soon join Lhwyd in Jesus College, becoming his closest friend and, perhaps, his only real confidant.[26] With John Wynne and William Anwyl (1668–c.1729), John Lloyd undertook a botanical expedition to Snowdon in atrocious weather in July 1686. Lloyd and Wynne sent Lhwyd accounts of their expedition: despite hardships and fear, it was a success, and they described the plants they had discovered.[27] Lhwyd respected their experience and knowledge and in August 1686 asked David Lloyd to assist him and John Lloyd in compiling a *hortus siccus* by collecting Welsh mountain plants, providing detailed directions on how to proceed (G, pp. 81–3).

The Museum's growing collections had to be organised. The first task was to compile the catalogues for the six Visitors as required by the Statutes. Plot and Lhwyd began work immediately and had completed all six catalogues by 1686, together with a consolidated single-volume catalogue for the Vice-Chancellor.[28] Lhwyd was responsible for three, perhaps four, of the inventories. By 1684 he had finished his initial task, the catalogue of the shell collection, for the Senior Proctor's Book, its title emphasising Lhwyd's official status as Underkeeper. Two other catalogues followed swiftly. The *Materia Medica* catalogue for the Professor of Medicine was ready by 1685, as was the Book of the Principal of Brasenose, a catalogue of zoological specimens. The latter has not survived but its content was included in the Consolidated Catalogue of 1695. The Book of the Junior Proctor was a catalogue, compiled by Lhwyd *c.*1685, of 'artificial works', man-made implements such as tools and weapons. In the Consolidated Catalogue this was replaced by an inventory of fossils, shells and minerals, mostly specimens from Plot's collection. Thus Lhwyd was responsible for cataloguing all the natural history collections of the Museum. Although the catalogues were inventories with little analysis or discussion, Lhwyd occasionally inserted comments based on his own observations. To the reference to 'Greater marine moss' in the *materia medica* list he added the comment, '*In Insula autem monensi copiosi nasci observavimus* [Moreover we have observed it springing up plentifully in the island of Anglesey].' A branch from a deformed ash tree reminded him of similar features on a willow '*in agro montis Gomeriei*'. This Latin rendering of 'Montgomeryshire' appears to have been one of Lhwyd's first (and not particularly convincing) attempts at onomastics.

Lhwyd found that providing descriptions of objects, especially those new to him, was a valuable discipline, and the extensive reading that the work required made him familiar with current scholarship in these areas. Before 1683 he had shown little interest in shells, and for his 1684 catalogue relied heavily on Lister's *Historiæ Animalium Angliæ* (1678).[29] He later explained to Lister how he had set about his task:

> I must confesse yt when I began this Catalogue, I was alltogether ignorant in ye Historry & Method of Shells; but having a good Collection at hand I first disposd them in such Method as seemd to me most agreeable to their Nature makeing all Shells

congenerous, wch I thought to agree in figure; & then consulted
all Authors yt had treated on yt Subject for every Species wch are
about 800. But I finde yt none besides yr Self … have hitherto dis-
tinguishd shells secundum genera et Species [according to genus
and species], much lesse reduced ym under method, as Plants &
ye other animals are done. (G, p. 89)

Fortunately, he was not dealing with a jumble of objects. The Tradescant
collection was already catalogued, and Lhwyd would also have been
guided by the organisation and labels of the *scrinia*, the cabinets and
drawers housing specimens such as Plot's collection of fossils. Plot's
work was to be an invaluable guide for Lhwyd when he embarked on
his own comprehensive descriptive catalogue of British fossils, since
Plot had created new standards of nomenclature, organisation and presentation for his collection, as well as providing excellent engravings
by Michael Burghers (*c*.1647/8–1727). Lhwyd may have found Plot's
mentorship irksome and came to disagree with him on many issues, notably on the origin of 'formed stones', the significance of certain fossils
and questions of description, but his work in compiling the catalogue
of shells and in dealing with the variety of objects listed in the second
version of the Book of the Junior Proctor profited from Plot's pioneering
efforts. Though Lhwyd found it difficult to acknowledge the intellectual
debt, working under Plot's supervision developed his cataloguing skills.

Despite his many talents and convivial nature, Plot was notoriously
ambitious and avaricious, Lhwyd justifiably calling him 'Dr Plutus' when
writing to Lister (G, p. 197). Plot soon become dissatisfied with his
relatively lowly position as Keeper and its poor rewards, and initially
attempted to supplement his salary by offering private chemistry courses
to paying pupils and selling an alchemical elixir.[30] During the second
half of the 1680s he ceased to teach regularly and sought additional
appointments elsewhere, becoming registrar of the court of chivalry
in 1687 and, very briefly, historiographer royal in 1688. His prolonged
absences disgusted Ashmole, who complained that Plot 'does no good in
his station, but totally neglects it, wandering abroad where he pleases'.[31]
Plot's frequent absences meant that Lhwyd acted as the public face of

the Museum, meeting visiting scholars and collectors and answering their queries. This prompted Lhwyd to suggest to the Philosophical Society in June 1686 that since

> many Curious Travellers when they visit the Repository, doe occa-
> sionally relate some remarques of their own experience, concerning
> things of *Nature & Antiquity*, he thought it might prove of some
> consequence to provide a Book that should lye in the Repository;
> wherein he might breifly set down, the contents of such relations.[32]

There is no record that this was done, but such discussions could encourage potential donors, the entries in the Ashmolean Book of Benefactors being a record of fruitful personal contacts.[33] Plot's absences also meant that Lhwyd had to acknowledge donations to the Museum. These formal contacts could lead to establishing a personal correspondence, as with Aubrey.[34] In the earliest surviving letter, 12 February 1687, Lhwyd acknowledged Aubrey's 'present' to the Museum and, as a Welsh-speaker, offered to assist him with his 'Queries'. Lhwyd and Aubrey were already acquainted, the letter referring to 'when we talked last together at Mr Wyldes' (G, pp. 133–4, date corrected from *EMLO*). One of the mysteries of Lhwyd's biography is the role of Aubrey's Shropshire patron, Edmund Wylde (1618–96).[35] It is not known how or when Lhwyd first met Wylde, but he greatly respected him as 'ye first yt gave me any encouragement to study British Antiquities; wherein haveing now got some small relish I think (how vain soever my endeavours may prove) I shall never quite forsake ym' (G, p. 206). Aubrey never trusted Plot but, as a firm friend of Lhwyd, eventually decided to bequeath his papers to the Museum rather than to the Bodleian library.

Some time before 1689 Lhwyd had commenced a correspondence with Martin Lister, whose generous but not altogether altruistic gifts to the Museum were discussed in Chapter 3.[36] Plot had initially acknowledged Lister's donations, but in the later 1680s Lhwyd took over that responsibility. Lhwyd knew exactly what the Museum required and was ready to make specific requests: 'if you can spare a sort of shell I have calld *Nerita longus purpurens* figurd 27A, be pleasd to send it down against ye Visitation' (G, p. 101). He was equally happy to respond to Lister's requests for fossil shells (G, pp. 114–15) and, later, for living land and freshwater snails, some found by Lhwyd himself or his friends, others gathered by local children, the specimens being packed in damp moss in strawberry baskets (G, pp. 203, 205–6, 208). Over the years the relationship between Lhwyd and Lister developed from being that between a junior curator and a valued benefactor to being a friendship of scholarly equals. Thus Lhwyd initially signed himself 'your most obliged and humble servant' but by 1690 had become 'your most affectionate & obliged servant' or 'your most affectionate friend and very humble servant'.

Plot and Lhwyd had planned to make a 'philosophical progresse' around Ireland in the spring of 1686.[37] Nothing came of this, but Lhwyd himself did travel there in the summer of 1688. On 3 July 1688 Henry Mordaunt, second Earl of Peterborough (1621–97), offered to finance a two-month collecting assignment (G, p. 122). Mordaunt's letter did not include the year, but this can be established by a letter from Bobart to Lhwyd 23 September 1688, where the former said 'our Ld P. who haeuing lost you for a while, wrot a somewhat severe letter to me' (G, p. 86). The arrangement – a horse, six pounds for expenses and instructions to collect specimens – provided Lhwyd with an opportunity to pursue his researches. Peterborough may also have been responsible for Lhwyd's decision to visit Ireland, for in another undated letter he enquired about 'a plant that growes in Ireland which is called Mackamboy' (G, p. 123). Mackinboy (variously spelled) is the Irish Spurge, *Euphorbia hyberna*, an element of the Lusitanian flora which contemporaries believed was confined to Ireland.[38] Scholars were interested in its medicinal uses as a violent purgative, so powerful that it allegedly worked when carried in one's pocket or if one had the misfortune to sit on a patch of it.[39]

In Dublin Lhwyd met members of the temporarily dormant Dublin Philosophical Society[40] such as Narcissus Marsh (1638–1713), Bishop of Ferns and Leighlin, a bibliophile, natural scientist and supporter of the use of Irish. He also met Dr Thomas Molyneux (1661–1733) and his brother William (1656–98), naturalists and enthusiastic enquirers in natural philosophy. In 1682 Thomas had published a broadside containing sixteen *Queries* for a description of Ireland intended for Moses Pitt's *English Atlas*. Pitt's bankruptcy put an end to that project, but Lhwyd obtained and preserved a copy of the questionnaire.[41] Lhwyd told his friend Richard Jones that Thomas Molyneux had amassed the 'best Collection of Books relating to Natural History' that Lhwyd had ever seen. Molyneux sought his assistance in a scheme for a natural history of Ireland, which Lhwyd was ready to embark upon the following spring, but his proviso 'if no disturbance intervene' proved to be all too prescient.[42]

Lhwyd described to Richard Jones his expedition to a small hill in Tipperary which, he had been assured in Dublin, 'produced all plants whatsoever that were Natives of Ireland … The hill grew more famous still as I went on in my journey, there being scarce any man or woman but had heard of it.' On finding the hill yielded no more than a few common plants, Lhwyd concluded, 'Soe much for the Irish Traditions which of all Nation's, come the nearest Perhaps to a Dream.' On the return journey to Dublin, Lhwyd came across a 'Popish Church', which he measured and carefully described: 'In shape it resembled a pedlar's stall … At the Alter end walld up, at the other open to the top, its Walls were of green clods, on the inside whereof grew allmost as many plants (spontaneously) as on the Hill above mentioned.'

On this visit Lhwyd first encountered Irish as a spoken language, his Dublin contacts having provided him with an Irish-speaking guide from Trinity College garden. His initial reaction was that

> The Irish \tongue/ hath many primitive words common with our language [Welsh], yet is as unintelligible to us, perhaps as Arabic. I belive its composd of old British, old Danish, Biscay, & perhaps \several/ others. Some of them told me that their Poets & such others as are expert in their language, have at least 20 words for every particular Notion. I have enclosd a Catalogue of \several/ words which I thought agreeable to some in our language. The Irish seem'd to me, when they spoke, to have the same tone that woemen have with us when they bewail anything.[43]

Here, in Lhwyd's (inadequate) analysis of the Irish lexicon and in the more soundly based 'catalogue' of some hundred cognate words, we find the beginnings of the study that would eventually lead to the *Glossography* (1707). The list is not a language-learning aid (as were his first lists of Irish words), but rather one of his earliest efforts in systematic etymology. Its motivation was phonetics, not dictionary definitions, and the intention was to show the relationship according to *sensum et sonum* (sense and sound) of words in the two languages.[44]

Lhwyd may have botanised in Anglesey on his return journey, since Bobart noted that, although his 'crossing over the water proved so unsuccessful', he had been 'very well recompted on our British coast' (G, p. 86). In Snowdonia, Lhwyd continued the work he had begun in 1682, drawing up a list of plants he had discovered (or revising his earlier notes) for his own use and for other plant-hunters visiting Llanberis. He sent the list to Tancred Robinson who, without consulting Lhwyd, sent it on to his close friend Ray, who was at this time completing his *Synopsis methodica Stirpium Britannicarum* (1690), a taxonomic catalogue of British flora which became the essential companion of field botanists for generations. Realising the value of the list, Ray immediately sent it on to his printer. Lhwyd was unhappy since, as he told Lister, the list was

> never intended for ye Presse; but onely to remain at Lhan Berys ... for the use of such as came thither a-simpling, & therefore much of it was writ in Welsh, wch is soe changed by often transscribing yt I scarce knew it to be that language when Mr Mod [Benjamin Motte, Ray's printer] shewed me Mr Ray's MS. One word being sometimes divided into 2 or 3, & elsewhere two or three words united; & so in ye 1st printed sheet I met wth a Greek word for Welsh. (G, pp. 89–90)

Although Robinson had attributed the list to 'an unnamed Welsh gentleman', Plot told Lhwyd on 10 June 1689 that Ray considered the catalogue to be the work of 'no triviall Herbarist, but a man of good skill in plants ... he suspects this catalogue to be yours.' He urged

Lhwyd to write to Ray 'as speedily as you can, for his book ... is in ye Press already' (G, p. 88). Before the end of June Lhwyd had sent Ray, via Lister, a collection of dried plants (G, p. 89). Although Lhwyd and Ray never met, and despite the great difference in age and experience, their correspondence is marked by their mutual respect and affection. Ray described Lhwyd in the Preface of the *Synopsis* as 'most expert not only in botany but in all of natural history',[45] an excessive compliment since Lhwyd still had much to learn.[46] Shortly before the *Synopsis* was published, Lhwyd told John Lloyd that it contained 'all my discoveries at Snowdon ye Summer 88, wch are above forty new plants' (G, p. 97). In the *Synopsis*, p. 7, Ray acknowledged Lhwyd as the source for plants in Snowdonia, as well as for seaweed such as twisted wrack '*in litore Monensi & Arvoniensis*' [on the shores of Anglesey and Caernarfonshire], a 'branched fistulous sponge' '*ad pagum Borth in agro Cereticense*' [in the Borth area in Cardiganshire], and, on land, '*Corallium minimum*' growing on limy rocks or cliffs at 'Lhan Derys [Llanberis], and Lhan Didno'.[47] As well as including these and additional Lhwyd discoveries in the second edition of the *Synopsis*, Ray had earlier noted several in his list of 'More rare Plants growing in Wales' included in the 1695 *Britannia*. Before the end of the 1690s Lhwyd was acknowledged to be the leading authority on Welsh alpine flora.

Lhwyd seems to have been particularly interested in two of his discoveries. What he considered to be *Subularia lacustris* is described in Latin on p. 210 of the 1690 *Synopsis* (illustrated on the facing page) and in English in *Britannia*, col. 702, as 'A Spindle-leav'd Water-Sengreen-like Plant [a plant resembling the Water-soldier, *Stratiotes aloides*], growing in the bottom of a small Lake near the top of Snowdon *hill, call'd* Phynon vrêch'. As Gunther maintained, the rather poor illustration clearly depicts *Isoetes* and not the plant now known as *Subularia aquatica*. Smith's *English Flora* gives Ray's *Subularia lacustris* as a synonym of *Isoetes lacustris*, and says 'Mr Lhwyd appears to have first remarked on it in Britain.'[48] Perhaps the illustration was omitted from the more generously illustrated third edition of the *Synopsis* because of its unsatisfactory nature. Lhwyd's other major discovery was described in Latin in the 1696 *Synopsis*, p. 233, and in English in *Britannia*, col. 702, as:

Bulbosa Alpina juncifolia pericarpio unico erecto in summo cau-
liculo dodrantali. *A certain Rush-leav'd bulbous Plant, having one
Seed-vessel on the top of an erect stalk about nine inches high.* On the
high rocks of Snowdon, *viz.* Trigvylchau y Clogwyn du ymhen y
Gluder, Clogwyn yr Ardhu Crib y Distilh, *&c. Mr.* Lhwyd. *It hath
three or four more narrow and short leaves upon the stalk.*

This is the Snowdon lily, which Lhwyd had not seen in flower. Since it
flowers from May to July, Lhwyd's discovery of the plant in seed may
have been made in August–September 1688. Bobart acknowledged
receipt of two boxes of plant treasures, 'there being diverse that we
never saw and perhaps never may againe', on 23 September 1688, but
was 'not soe fortunate as to find that bulb wch in yr letter you direct
me to [possibly the Snowdon lily] at the top of the biggest box', and
feared that it had not been included (G, pp. 85–6). A few years later,
in an undated letter, possibly written in May 1695, Lhwyd told John
Lloyd that 'Dr Foulks'[49] had

> found a plant in flower in Snowdon, which I have mention'd in
> Mr Ray's *Synopsis*, but with the Addition that I never saw the
> flower of it. I suppose 'tis either the *Subularia* ... or the *Bulbosa
> Alpina juncifolia*, but would gladly be inform'd whether of them ...
> and would be much oblig'd to him for the best Description he can
> give of the flower. (G, pp. 273–4)

When Lhwyd visited Snowdon in 1696 he told Robinson that he saw
'the litle Bulb ... plentifully in flower', but as for the *Subularia*, 'the Lake
where it grows is so high, that men have seldom occasion to come near
it, so that I have but slender hopes of any account of it's flowering' (G,
p. 308). He also sent Ray a specimen of 'the Bulb with a single flower,
wch [Lhwyd] had seen in seed before'. It puzzled the recipient, 'if it
be not a plant *sui generis* ... it may be referred to *Ornithogalum*.'[50] The
Snowdon lily, or Mountain Spiderwort, *Brwynddail y Mynydd* (rushy
leaves of the mountain) was named *Lloydia alpina* by R. A. Salisbury
in 1812: 'As it constitutes a distinct genus, I have named it after the

celebrated EDWARD LLHWYD, Esq. who communicated so many scarce plants to RAY.'[51] After Salisbury's faulty nomenclature was corrected by Reichenbach in 1830, it was known as *Lloydia serotina*.[52] However, DNA sequencing has recently led to all species of Lloydia being included in the genus *Gagea*, so that it is now known as *Gagea serotina*. Although found in Asia, Europe and North America, it is a very rare plant in Wales. Over-collected in the past, it has been legally protected since 1975 but is restricted to some five sites in Snowdonia and remains extremely vulnerable to climate change.[53]

Ray and Lhwyd soon began discussing fossils and exchanging specimens, Lhwyd sending Ray duplicates from the Museum's collections, a practice authorised in Ashmole's Statutes:

> I only expect your commands for some figured stones. Those that this country [Oxfordshire] affords are chiefly in imitation of shells. We have none that resemble fish, or any other animals besides, nor that have the resemblance of any plants. *Cornu Hammonis, Asteriscus, Asteria s. Astroites*, and *Belemnites* of divers sorts, we have plentifully, as also some others that I cannot compare to any natural bodies that I have any notion of. One quarry within two miles of Oxford [possibly Cowley] I have searched at least forty times, and sometimes had five or six with me; yet last Saturday I discovered there three varieties of *Glossopetræ*, though none had ever been observed in this part of England before, for what I can learn. (G, p. 100)

Lhwyd's letters to Ray and Lister reveal his growing interest in fossils as objects and, as he came to realise the wealth and diversity of material in local quarries and gravel pits, prompted questions about their context, distribution and origin. Lhwyd's detailed comments to Lister in August 1689 accompanying a gift of fossils 'all gathered within three miles of Oxford' reveal he had come to understand the need for a more systematic study:

> I must confesse yt in collecting these Stones, I was not sufficiently curious in any Observations about them; however I have noted,

1 That some Stonepits near Oxford affoard at least 30 sorts of form'd stones; whereof most are Cochlites.

2 That ye same Species are to be found at ye same time & place in several degrees of magnitude: thus I have seen 15 \several sizes/ of ye *Echinites rotelis* I have sent you; & soe of some other stones, exactly as in shells.

3 Of such qarries as I have hitherto seen; those that consist of a Sandie stone, affoard ye greatest variety of *Cochlites*.

4 I have seen some Stones of yt kinde I call *Echinites laticlavius* \(degener.)/ *abortivus* (as I suspected) *in fieri* [in process]; whence I conjecture that they never were entire Echinite's & casually broken; but formd as we finde them; wch is in ye figure. I have sent them to you; & tho I found hundreds of them; yet I never met with any otherwise. (G, pp. 90–1)

As Lhwyd compared his finds with published accounts, his descriptions grew increasingly detailed. Since some of his discoveries were hitherto unrecorded, he was obliged to create names for them. Outlining his views on naming 'Natural bodies' to Lister, he argued that although short titles were the 'most serviceable' they did not 'allways distinguish the Species from those that are congenerous with them; or expresse all their properties'. On the other hand, since 'descriptive' titles, 'such as are composed of four or five & so to 8 words' could be unwieldy, he judged it convenient 'to give them short titles, leaving the rest of their properties to their figures, descriptions, &c.', or else employ two titles, a short 'nominal' one, the other 'descriptory, longer, when reqisite' (G, p. 248). Such descriptive names would be the norm until the advent of Linnaean binomial nomenclature in the mid-eighteenth century.

Lhwyd discussed with his correspondents questions that were beginning to intrigue him as he observed the great variety of fossil forms. Some, such as bivalves and other molluscs, were similar to living organisms and consequently were what Rudwick calls 'easy' fossils, but others, such as belemnites, were unlike anything Lhwyd had encountered and would thus fall into Rudwick's category of 'difficult' fossils.[54] Writing to Ray in November 1690, Lhwyd set out some of his thoughts on fossil origins:

Whether they were ever the tegumenta of animals or are only primary productions of nature in imitation of them, I am constrained ... to confess I find in myself no sufficient ability or confidence to maintain either opinion, though I incline much to the latter ... On the one hand, it seems strange if these things are not shells petrified, whence it proceeds that we find such great variety of them so very like shells in shape and magnitude, and some of them in colour, weight, and consistence; and not only resemblances of sea shells should be found, but also of the bones and teeth of divers sea fish, and that we only find the resemblances of such bodies as are in their own nature of a stone-like substance. On the other hand, it seems as remarkable that we seldom or never find any resemblance of horns, teeth, or bones of land animals, or of birds, which might be apt to petrify, if we respect their consistence; insomuch that I suspect few formed stones are found (at leastwise in England), except in some extraordinary petrifying earth, but what a skillful naturalist may, and that perhaps deservedly, assimilate to some marine bodies. (G, pp. 110–11)

In December 1689 Lhwyd accompanied Plot on an excursion to Kent 'yt we may try what discoveries we can make there in Natural History' (G, p. 95). Lhwyd probably obtained some fossils there, since Plot told him in January 1691 that 'the smith of Borden has not forgot you, having gotten several *Echinites* for you' (G, p. 133). What was of more lasting significance for Lhwyd was that this journey was part of Plot's scheme 'of encompassing Britain, along ye Seacoast, & of writing a Natural History in Latin, under the title of *Zodiacus Britanniæ*' (G, p. 97, amended from *EMLO*). Some fifteen years earlier, Plot had sent Dr John Fell the 'Design ... [of] a Journey through *England* and *Wales*, for the Promotion of Learning and Trade', which described in detail enquiries on a wide range of 'curiosities' in natural history and antiquities and sought Fell's encouragement for the ten-year project.[55] Lhwyd would recall this plan as he developed his own ideas on a natural history of Wales and the financial support it would require.

In June 1690 Lhwyd told Ray he had viewed for the first time Cole's collection.[56] Plot did not know about the journey, Lhwyd warning Lister, 'Be Pleasd to say nothing of this journey to our ffr[iend] Dr P. least he finde fault with my absence so longe from ye musæum' (G, p. 107). Lhwyd 'received abundant satisfaction by my journey' as Cole

> received us, though all unknown to him, very friendly, and spent six hours in showing us his collection, without any interruption, or the least sign of being weary. It consists altogether of natural things, and seemed to us a very extraordinary collection for one person (and who, perhaps, had not the advantage of a liberal education to invite him to such studies) to be able to amass together. (G, p. 104)

Since the collection recorded where each specimen had been found, and did not duplicate material in the Ashmolean, Lhwyd was later to spend much time attempting to acquire it for the Museum.

Within six years Lhwyd had established himself in the scientific community. His cataloguing duties at the Ashmolean had provided him with an invaluable training and grounding in scientific literature, and his supervision of visitors had made him known to visiting scholars. His reputation for discovering rare plants and, increasingly, fossils, had enabled him to establish a correspondence with three of the most eminent natural historians in England, Ray, Lister and Robinson.[57]

KEEPER OF
THE ASHMOLEAN MUSEUM

Although Lhwyd was now an established scholar, he found his insecure status as Plot's assistant irksome and sought to travel further afield. In the summer of 1690 he heard that the 'Botanic Club in London'[1] had 'entertain'd some thoughts of sending me to the Canarie Ilands to make what discoveries I can in plants' (G, p. 107). He set about learning Spanish, but nothing came of the scheme. Even after his appointment as Keeper, Lhwyd indulged in hopes of foreign travel, perhaps to escape his uncongenial administrative responsibilities. Writing to Lister in September 1692, he speculated whether the Visitors would allow him to take up a temporary post in Barbados (G, p. 167). Before the end of the year there was more talk of an expedition to the West Indies.[2] 'I could wish 'twere rather to ye Canaries, but beggars must not be choosers,' he told Lister in late December 1692 (G, p. 173). Friends such as Lister[3] and Ray tried to dissuade him from venturing: 'it is so hazardous both for the danger of the Sea, & at this time of the publike Enemy, & the great change of air and diet.'[4] By mid-February 1693 Lhwyd had accepted that the West Indies scheme was dead, and considered spending 'one or two Summers in Germany' to 'capacitate my self to be a Lecturer in minerals, (& perhaps Coyns & Antiquityes) in ye Museum' (G, p. 175, date corrected from *EMLO*). He finally abandoned the idea of visiting the West Indies after his 'intimate friend' Robert Parry (c.1667–93?)[5] died on arriving there; as Lhwyd told John Lloyd in May 1693: 'My intended voyage to ye West Indies is quite laid aside:

poor Robin Parry, went as minister to ye Barbadoes, and died the third day after he landed' (G, p. 191).

Lhwyd's status and finances in the early 1690s were precarious. Despite his title of 'Underkeeper', in reality he had no official position and Higgins was his senior. Plot had to divide the inadequate income deriving from museum fees as best he could, paying his own salary and that of Higgins and allowing Lhwyd what was 'convenient'. In January 1691 Lhwyd complained to Lister that 'in reqitall of my attendance at ye Musæum this last year' Plot had allowed him 'six pounds seven shillins & two *Historys of Staffordshire*' (G, p. 131). In the summer of 1689 Plot began to speak openly of retiring. This increased Lhwyd's sense of insecurity, since the appointment was Ashmole's prerogative and it was known that he favoured his nephew George Smalridge (1662–1719), whose education he had financed. Lhwyd thought Smalridge unsuitable since he had no interest in natural science, 'having as great aversion to that sort of Study, as I may have inclination to it; & as ignorant yt way, as he's otherwise ingenious' (G, p. 92). Ashmole might, however, compel Smalridge to take up the post by threatening to cease financing his studies were he to reject the offer. As Lhwyd became increasingly anxious, he began to doubt everyone's motives and intentions. He was reassured by Lister, who regarded him as Plot's natural successor, as did Plot himself. Mid-August 1689 Lhwyd sent Lister arguments to counter Ashmole's reasons for refusing to appoint him, the most important being that Ashmole's 'great & most laudable design of promoteing ye knowledge of Nature, will be utterly over thrown if he prefers one … who being ignorant in \such/ studies, can never instruct any others therein' (G, pp. 92–3). Lister's reply, in an undated letter postmarked 20 August, mentioned a fresh problem. Plot had just informed him that upon Lhwyd 'entering into my L. Petrboroughs service, an other was put in yor place; who still enjoys it, as under keeper. this makes your case worse. for I did verilie believe you had been in possession of it.'[6]

That difficulty was resolved when Higgins resigned in September 1690. The following month, Plot resigned his chair of chemistry but remained Keeper. As Lhwyd grew increasingly impatient with Plot's apparent indecision (in fact, Plot may have delayed his resignation to assist him), he considered writing directly to Ashmole to press his claims. Fortunately he asked Lister's advice as to 'what sort of Letters will be most agreeable to his humour; for altho I may præsume so far on my own judgement, yt I shall not disoblige him in what I write' (G, p. 109). Lhwyd failed to realise that the views on the nature of the Keepership expressed in this letter would have been wholly unaccept-able to Ashmole:

> I finde he [Ashmole] reqires constant attendance at the Musæum, but constant Enqiries after natural Productions would be farre more usefull towards the advancement of knowledge; & we may reasonably suspect; yt when any one shall hereafter endow it with a travailing Fellowship, it may lose ye title *Ashmoleanum.* (G, p. 109)

Had Ashmole read these words, he would undoubtedly have rejected Lhwyd as a candidate for appointment as Keeper, but they indicate clearly Lhwyd's vision of the Museum as a research institution ground-ing its 'enquiries' in fieldwork.

Plot finally resigned the Keepership in December 1690, 'much against the grain', commented Lhwyd, and in January 1691 Lhwyd and Plot visited the Vice-Chancellor to confirm the transfer of the Keepership and the keys of the Museum. Lhwyd described the events in a letter to Lister 17 January 1691 that is remarkable for Lhwyd's expression of the suspicion and ill will he harboured towards Plot who, as far as one can judge, had always shown a friendly affec-tion for him:

> Hence I conclude yt I am wholly indepted to yr goodnesse for this happy deliverance out of his clutches; for (to give him his due) I think he's a man of as bad Morals as ever took a doctors degree. I wish his wife a good bargain of him; & to my self yt I may never meet with ye like again. (G, p. 131)

Lhwyd had shared his long-standing hostility with his intimate friends; in July 1686 William Anwyl mentioned 'the perfidious Dr', and gloated over his 'ill successe in his last history'.[7] Plot was unaware of Lhwyd's feelings, sending him a friendly letter at the end of January: 'Pray let me hear from you sometimes how the Musæum and Natural History thrives, as you shall from me, for tho' the London and Oxford Societies sleep, yet let *us* be awake.' The letter closes: 'I am dear Ned, Thy most faithfull Friend, Rob. Plot' (G, p. 133). When Plot became Secretary of the Royal Society for a second time in November 1692, he offered to print Lhwyd's contributions 'without charges' (G, p. 170), a generous offer which Lhwyd commented on ungraciously (G, pp. 172–3). Although respected and well liked by his peers and contemporaries, Lhwyd seems to have been personally insecure and consequently all too ready to doubt the sincerity and motives of those who professed to support him, a trait perhaps inherited from his father and accentuated by his difficult upbringing.

During the year-long wait for the keepership Lhwyd moved into new areas of interest as he began to explore questions of phonetics and orthography. The etymology of Montgomery referred to in Chapter 4 was an unconvincing guess, but from about 1689 onwards queries from friends and colleagues and his own inclinations led him to develop a more structured approach to the study of antiquities, place-names and language. The 1688 glossary of comparative phonology has already been noted in Chapter 4. Around this time too he began to use 'q' for 'qu', on the grounds that 'qu' did not represent /k/ followed by /w/ but rather a single phoneme containing a rounded element [kʷ] that should logically be represented by a single grapheme, 'q' rendering 'u' superfluous. Plot disliked Lhwyd's new style, especially when it was used in the appreciation of his own gifts and service in the Museum's Book of Benefactors:

> I [do not] like your writing the Q's without U's after them, for though it be sure that the Letter U is always included in Q, yet it being a singularity and savouring of affectedness (if I might advise) I would not have you use it at all, lest you incur censure; however let it alone in this of myne and doe what you please for the future. (G, p. 134)

Lhwyd also followed this principle in English between 1689 and 1692, writing *qarries, qestioning, acqaint.*

The etymology of Welsh words also occupied his mind during these years. Writing to Plot in May 1690[8] he countered an assertion by Edward Bernard[9] in his *Etymologicon Britannicum* (1689)[10] that since about half the Welsh words in Dr John Davies's *Dictionarium Duplex* (1632) derived from Latin and a quarter from English, no more than a quarter were of native origin. Lhwyd wrote under a pseudonym, 'Meredydh Owen', in deference to Bernard's scholarly standing for, as he later explained to Lister, 'Dr Bernard being an eminent critic, I thought it not convenient to subscribe my own [name]' (G, p. 173). It was a futile artifice, since no other contemporary Welsh etymologist could have produced such arguments. Lhwyd attempted to devise a methodology to determine which words had been borrowed from English or Latin and which were 'cooriginall':

> a few Welch wordes that doe indeed agree with the English both in sound & sense, & yet could not probably be receivd into our language from the English Conquerours ... in regard they are for the most to be found in the *Amorican Lexicon* ... And the Brittains, who went hence to Amorica, left us in the year 384 whereas the Saxons came not in till the year 450.[11]

Welsh words could be shown to be cognates and not borrowings if they were also to be found in languages whose speakers historically had not been in contact, as the Irish and the Romans or the Bretons and the English.[12] Six years later Lhwyd's friend Humphrey Foulkes (1673–1737) carried out a similar exercise with a list of French and Welsh words agreeing in sound and significance; a comparison of Latin, French and Welsh, with 'ye help of your [Lhwyd's] Irish' as an outside control, should reveal elements of the vocabulary of Gaulish, since those French words 'having no affinity with ye Latin must be of Celtic originall'.[13]

Lhwyd's interest in language contacts was reinforced when Ray told him in November 1690 that he was preparing a second edition of his *A Collection of English Words, not Generally used, with their Significations and Original, in two Alphabetical Catalogues, the one of such as are proper to the Northern, the other to the Southern Counties,* first published in 1674.[14] Lhwyd's well-worn copy of the book (Ashmole A1) contains numerous notes and additions in his hand, inserted at various times during his tours, testimony of his regular use of the work. His enthusiastic response to Ray's projected second edition is a further indication of his sustained analytical thinking about language contacts. He considered that some English north-country words that 'bear affinity with the Welsh, both in sound and signification ... possibly may be some remains of the British tongue continued still in the mountainous parts of the north', but recognised the difficulty of distinguishing between cognates (shared 'affinity') and borrowing (G, p. 110). He resolved matters by using the same criteria of historical language contact utilised in his 'Meredydh Owen' letter.

Ray published the second edition of his *Collection* in 1691, 'augmented with many hundreds of Words, Observations, Letters, &c.'[15] In the preface he thanked contributors, including 'Mr. Edward Lloyd' who had communicated a list of '*British words* parallel to some of the Northern Words in this Collection, from which probably the Northern might be derived'.[16] 'A Catalogue of Local Words parallel'd with *British* or *Welsh*, by my learned and ingenious Friend Mr. *Edward Lloyd* of *Oxford*', occupying pages 122–30, listed forty numbered words in two columns headed *English* and *British*. The 'British' list notes possible Welsh (and occasionally Breton or Irish) parallels, with comments on their origin. Some were noted as local or dialect words, e.g. No. 15 '*dwbler Cardiganshire*, signifies the same' [as English 'doubler', a dish]. A few entries, such as No. 23, are more elaborate: '*hilio*, to cover. Perhaps we have received it from the *English*, which may be the reason Dr. *Davies* hath omitted it in his Lexicon. It is a word generally used in North *Wales*.'

A shared interest in natural history, antiquities and etymology led to Lhwyd establishing a correspondence from June 1691 onwards with the eminent scholar William Nicolson (1658–1727),[17] who was compiling a history of Northumbria. They initially compared north Wales and Cumberland, Nicolson noting in March 1692: 'Our Countey is mostly (as you rightly guess) of the same nature wth North-Wales. Onely, I think we have greater store of Mettals; & therfore, it may be, fewer form'd Stones are to be mett wth in our Mountains.'[18] He was particularly interested in place-names:

> A great many of our Towns (as Caerlile, Penrith, Penruddoc, Caerdornoc, &c) with most of the names of our Rivers and Mountains are Brittish. I shall send you a large Vocabulary of 'em; since you allow me the confidence to hope that you will readily explain them for me.

Acknowledging a 'choice present of form'd stones' from Lhwyd in October 1692, he thanked him especially for 'the excellent Rules ... for ye derivation of ye Brittish names of our Rivers', adding that he had been 'scrapeing together the names of our Mountains ... But I do not find so many of them to be Brittish as I expected ... and some few yt still retain ym have them vilely corrupted.'[19]

Mid-May 1693, Lhwyd asked John Lloyd to gather information to assist Nicolson's researches. He requested 'a catalogue in your next of all ye ancient towns, castles, & forts' in Merioneth and Denbighshire: 'I desire chiefly ye names of such ancient (and at present mean) places as are not mention'd in ye maps', though those that were included 'are so false written, that onely those as know ye countrey very well can understand them'. He would also welcome 'a catalogue ... of the mountains and lakes of Meirionydhshire, with yr brother David's interpretation and glosses on them'. Lhwyd continued, 'My design in this, is partly to observe ye method our ancestors used in nameing places; and partly to gratify a very ingenious gentleman ... viz. Mr William Nicolson' (G, pp. 191–2).

Following his appointment as Keeper in January 1691, Lhwyd's distaste for administration led to a marked deterioration in the quality of account keeping.[20] He clearly did not believe that the Keeper needed to be present constantly at the Museum and was content to leave a deputy in charge while he undertook field trips. In April 1691 he was able to enjoy an expenses-paid excursion. 'Two Danish Gentlemen[21] yt are great lovers of these sort of Enqiries' paid him a guinea each for collections of fossils. Lhwyd cynically remarked that 'I afforded them a very good bargain to encourage \them/ ... [they] have now such a relish of that sort of Diversion yt I suppose they'l never qit it.' His investment paid off, since they asked him to act as a guide, offering 'to bear my charges if I goe with them' (G, p. 137). There were so many places to see that 'allmost all our time was taken up in riding', but unfortunately one of the Danes 'had never been on horseback before, & proved so bad a Horseman, yt we could advance but 20 miles aday', prompting Lhwyd to complain, 'had I spent so many days alone, I doubt not but I had discovered ten times as much as we all did, now.' During a nine-day tour they visted Cirencester, where they bought Roman coins, Woodchester, where they viewed a Roman mosaic pavement, and coal pits at Acton, where they found fossil plants. At Bristol they met Cole, who subjected them to a two-hour reading of 'part of a MS of Equivocal generation' before allowing them to view his collection (G, pp. 138–9). In Bath they visited Cole's friend John Beaumont (c.1640–1731).[22] A doctor, best known today for his bizarre accounts of his dealings with the spirit world, visions which his acquaintances unkindly ascribed to his fondness for drink,[23] Beaumont was an enthusiastic collector of fossils who built up a notable collection of specimens from the Mendip lead mines and pioneered the exploration of local caves, notably Wookey Hole. In 1685 he had issued a prospectus for a natural history of Somerset on Plotian lines, but the scheme lapsed following the Monmouth rebellion.[24] In October 1694, as Beaumont became more interested in the spirit world than in fossils, he attempted to sell his collection to Oxford University for twenty pounds, an excessive price in Lhwyd's opinion for a collection which largely duplicated the Museum's holdings (G, p. 254). On the return journey they found 'some Cutbert's beads & some Fluors' in the Mendip lead mines, before travelling via Wells to Wookey Hole. On the penultimate day they visited Stonehenge. Lhwyd was so impressed by the size of the stones that, when telling Ray some months later about 'prodigious heaps of stones' on the summits of the highest Welsh mountains, he said that 'many of them [were] of the largeness of those of Stonehenge' (G, p. 159).

Shortly after his return, Lhwyd 'at last perform'd the journey to Huntingdon' seeking a hill reputed to contain many shells. No such hill existed, he told Lister on 26 May, Huntingdonshire 'being a low clayie Countrey', but the return through Northamptonshire was more fruitful:

> I have found perhaps as many Figured Stones, as that hill wherever it be, may afford ... This last County ... affoards considerable variety of figured stones. I have several Bivalves from thence & other places, of this journeys Collecting, which shall be sent you e're long. (G, p. 144)

He managed to spend a few days in August fossil hunting with the compulsive collector James Petiver (c.1665–1718).[25] In a recently rediscovered travel diary, Petiver records spending five days with 'the Ingenious Mr Floyd ... we went about Thirty miles into the Country a lithoscoping',[26] perhaps as far as the sand and gravel pits of Faringdon (Berkshire, now Oxfordshire).[27] Lhwyd complained to Lister on 25 August that preparing for the Visitation had allowed him 'no time to go abroad to persue my Enqiry's'; he now intended spending a week at Cirencester and a week at Gloucester, 'leaving a Friend in ye mean while at ye Museum' (G. p. 148). All these expeditions are further evidence that Lhwyd considered that his primary responsibility as Keeper was to collect specimens. The importance of safeguarding the collections was soon to be brought to his attention.

Between 17 and 21 September 1691 the Museum suffered a serious robbery. Although Ashmole's Statutes had laid out rules to safeguard exhibits, once visitors had obtained permission to view the collections an atmosphere of trust appears to have prevailed. The robbery was a planned theft of at least thirty treasures, mostly gold and silver coins, medals, rings and pictures, including a miniature of Aubrey by Samuel Cooper, recently donated by Aubrey himself. Lhwyd was extremely distressed by the loss and was especially fearful of Ashmole's reaction and the consequent threat to his own position. In an attempt to discover the thief he turned to two trusted London friends, Thomas Madox (1666–1727), the great legal antiquary,[28] and William Charleton (1642–1702), an eminent naturalist and collector who was familiar with London dealers in curiosities.[29] Lhwyd sent Charleton a list of some of the missing items, hoping that dealers – he suggested several names – would recognise any that might be offered to them. His own suspicions had fallen on 'a foreign gent. supposed by his speech to be a German, but speaks tolerable good English & Latine'.[30] Charleton, whose own collection had suffered from pilfering, expressed his sympathy: 'Losses of this kind will happen to all persons who expose things of this nature to publick view Let them be never so vigilant', and passed the list of the stolen goods 'to particular friends of mine, who will manage the Concern with privacy & fidelity'.[31] One of those 'friends' was Hadriaan Beverland (1650–1716), a classical scholar whose writings, obsessed with sexuality in antiquity, had led to his expulsion from the University of Leiden. Following a period of imprisonment, he emigrated in 1680 to England, where he eventually worked his way into the orbit of Hans Sloane (1660–1753).[32] Described by Ovenell as one 'who appeared to know so well the less reputable channels of London dealers',[33] by 1700 Beverland had succumbed to paranoia – Hearne later said he was 'in a craz'd condition'[34] – and produced a strange document which strongly hints at his complicity in the theft.[35] Despite evidence that the thief might have been Dutch, Lhwyd never expressed any suspicion of Beverland. He followed the trail to London,[36] the University granting him the substantial sum of four pounds and ten shillings, over £730 in present-day value, for his expenses.[37] He also had enquiries made in the Netherlands, but none of the items ever came to light. In December 1691, uncertain whether Ashmole had heard of the theft, he wrote to Lister, '"Tis discoursed here yt Dr Plot has told him of it, how truly I know not' (G, p. 152). But by early February 1692 Lister had smoothed things over, 'making my excuses with Mr Ashmole'. Charlett had shown Lhwyd a letter from Plot saying that 'he was not the man that acquainted Mr Ashmole with our mischance'. When Ashmole had asked Plot about the theft, Plot told him that

it was a thing he had been afraid of these many years; & \had/ spoken with all the V. Chancellors that had been, while he was Keeper; about locks for preventing it. Adding that he has a greater kindness for me, than to possesse Mr Ashmole with a mistrust of my fidelity or care. (G, p. 155)

Since Ashmole was now terminally ill (he died mid-May 1692), his debility may explain his uncharacteristically feeble reaction. Ashmole bequeathed about eleven hundred of his printed books and some six hundred manuscripts to the Museum, the remainder (including books which would have been of use to the Museum) being auctioned in 1694.[38] When the books arrived in the summer of 1692 Lhwyd told Lister that it would take him at least six or eight months to catalogue them.[39] In February 1693 he complained to Lister about 'a task yt's to me neither pleasant nor profitable' since it diverted his efforts from his work on fossils (G, p. 175) and, as he told him in April, he considered that the astrological texts were useless:

> You take care to send us nothing but what is valuable & perti-
> nent. But I could heartily wish Mr Ashmole had also done the
> same in his Legacy of Books; & instead of many MS volumes of
> Mr Napeir's Astrological Practice in Physic,[40] & above five hun-
> dred other Astrological Books; I wish he had given us 50 of his
> best Books relateing to coyns and other Antiquities, & to Natural
> Philosophy. (G, p. 177)

Lhwyd recognised that Ashmole's 'MSS. relateing to Heraldry & his collection of Pamphlets & English Poems' were of value, but in early February 1694, he reiterated his opinion that 'a great part of [them] are uselesse to learning, & those are of judiciary Astrology; and some old Monkish, Chymistry & Physic. This I shall take the Liberty of passing over slightly; but shall take more pains with the rest.'[41]

To assist him in cataloguing, by December 1691 Lhwyd had appointed a young Welsh undergraduate, William Jones (d. 1702), perhaps as Underkeeper.[42] Jones may have been a member of Lhwyd's circle since the 1680s and became a trusted assistant in his travels and researches. Although the Catalogue of Ashmole's library, 'above a hundred sheets', was 'allmost finish'd' by May 1693 (G, p. 192), the

'Slavish Employment' was not completed until the end of 1694.[43] Lhwyd entrusted Jones with cataloguing the uncongenial astrological books. Another large bequest soon followed, the library of Anthony à Wood, who died in November 1695. According to Hearne, Wood left his books to the Museum rather than to the Bodleian because 'Mr Lhuyd always shew'd him a great deal of Civility, & would readily produce him any Book when he came to the Muséum.'[44] Lhwyd catalogued Wood's manuscripts himself, allocating the printed books to the Under-librarian Robert Thomas (b. 1681).[45] Like many contemporary libraries, the Museum used a marked-up interleaved copy of the 1674 Bodleian printed catalogue (from Wood's donation) as the basis of its own listing.[46] Other gifts continued to arrive and had to be processed, and Lhwyd was obliged to spend time cultivating benefactors, but throughout these years his main concern was to expand and arrange the fossil collections.

The robbery failed to deter Lhwyd from travelling, as he now combined his quest for the thief with searching for new fossils. In November 1691 he went to Gloucestershire

> partly to enquire after ye Dutchman I suspected ... & partly to finish ye collection which I design for ye subject of ye *Lithol. Oxon*. ... I find that ye best place I have hitherto seen for variety of Formd Stones is the Severn shore about Frethern[47] & Purton Passage & doubtlesse many other stonie places on both sides ye river betw. Glocester & Bristol. (G, p. 150)

He also continued his botanical studies. In December 1691, for example, after acknowledging gifts of books from Richardson and identifying plants received from him, Lhwyd told him about recent botanical discoveries by Ray and his friend and assistant Samuel Dale (1659–1739) (G, p. 151). Fossils were now, however, Lhwyd's main interest. Between August 1691 and February 1692 he sent Lister enthusiastic reports of his discoveries in a gravel pit at Faringdon,[48] and of fossil oysters at Rangewell Hill,[49] Headley, south of Epsom, where 'I dug out some my self with a Mattok' (G, p. 155).

Lhwyd's collecting during these years led to his first publication discussing fossils, a letter to Hemmer describing and illustrating recent finds, which appeared in the May 1693 number of the *Philosophical Transactions* (G, pp. 178–82). Lhwyd believed that these were fish fossils, but we now know that some were the remains of ichthyosaurs, creatures which Lhwyd could not have envisaged, but had discovered well over a century before they were recognised in 1811.[50]

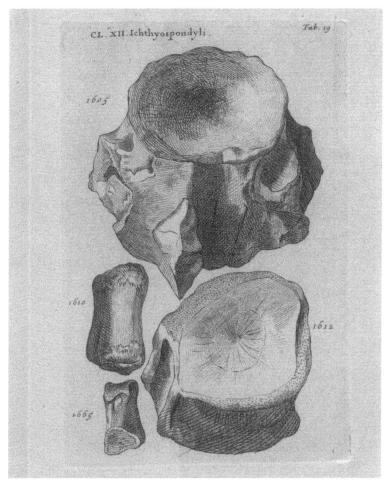

FIGURE 2 *Ichthyospondyli*. Some of Lhwyd's 'fish' fossils were later recognised to be ichthyosaur vertebrae

FOSSILS AND 'FORMED STONES': *LITHOPHYLACII BRITANNICI ICHNOGRAPHIA,* 1699

L hwyd's enthusiasm for fossil hunting was shared by several scholars in the 1690s, the most prominent being John Woodward (1665/8–1728), with whom he was to have a troubled relationship.[1] As a teenager Woodward was taken up by Peter Barwick (1619–1705), physician to Charles II, who provided him with medical training and supported his successful application for the professorship of Physick at Gresham College. Woodward told Lhwyd in April 1693 that the post was worth 'but £50 a year each, besides Lodgings: mine are very good; I lett part of them for £30 a year.'[2] In other words, he enjoyed a more substantial and secure income than Lhwyd and could expect to supplement it by his medical practice. Woodward collected his first mineralogical specimen in 1688 and his first fossil, a Jurassic brachiopod, in the Cotswolds on 13 January 1690 – Woodward always noted the date and location of his finds. Furthermore, he died a wealthy man, leaving a bequest of a hundred pounds a year to the University of Cambridge to endow a 'Professorship of Fossils', now the Woodwardian Professorship of Geology. Since the collection of almost ten thousand specimens had to be audited every year, it has survived intact in its original five display cabinets and forms the basis of the present Sedgwick Museum. Lhwyd, who died in debt, could make no such provision, and his collection was dispersed and largely lost following his death. Woodward also enjoyed public recognition far earlier than did Lhwyd, being elected a Fellow of the Royal Society in 1693, whereas Lhwyd did not become an FRS until 1708.

Lhwyd first met Woodward sometime after the end of April 1690, when Woodward was recommended to him by William Charleton as 'a very ingenious Gentleman'.[3] Lhwyd told Lister in July that Woodward 'seems to have made a wonderfull progresse (considering his age) in several sorts of Knowledge', but expressed some reservations:

> he does not much qestion but he has found out ye Causes of those Productions [fossils]; & added that they seem soe plain yt he wonders no body thought of it sooner. What notions he has therein I know not; but it seems qestionable whether he be sufficiently experienced in such *Observations*, as to be able to satisfie mens Curiosities in soe nice a phenomenon. (G, pp. 106–7)

Unfortunately only one of Lhwyd's letters to Woodward has survived. In his earliest letter to Lhwyd, dating from late October 1691, Woodward mentioned meeting him in London and his disappointment that they had failed to meet again at 'the Rainbow Coffee-house in Fleet-street'. After noting good places for fossil hunting, he offered to assist Lhwyd's researches: 'what observations I have made you may freely command; & wherever I have duplicates of stones &c which you have not, they are at your service.'[4] They exchanged thoughts and specimens amicably during the early months of 1692, but towards the end of March Woodward wrote:

> I come now to a passage in your Letter which I confess I could not have expected, you urge 3 or 4 Arguments that imply I am ungratefull in diminishing the just value \of the things/ you presented me, & that I would insinuate untruely that I had them before, when I had them not ... I again assure you that I have now actually by me, & ready to shew to you ... specimens of every sort you presented me, except the Crabs claw, which I never pretended

too … 'tis not impossible but I might use the term (reserved) finding so fair & remarkeable a stone amongst those you sent to Mr Odhelius,[5] & recollecting that (I ha\ving/ wrote to you about two months before to send me a list of such plants as you had observed in stone) you never in the least mentioned this plant.[6]

For his part, Lhwyd told Lister in mid-April:

I have some correspondence of late with your neighbour Mr Wod. but I belive it will not continue long; for I suspect him to be too cunning for me; & not over communicative of his things or Notions; nor liberal of his good character, to such as may deserve it.[7]

Nevertheless they continued to correspond and to meet. At the end of August Woodward told Lhwyd with his customary assurance: 'You may please to remember that when you was here [London] we were discoursing of some passages in Dr Plots … *Oxfordshire*, but were interrupted by the shortness of your time: I am sure the Dr is mistaken in some, & I think in all, of them.'[8]

Woodward's next letter, dated 20 October, was very different in tone. Lhwyd had unwisely expressed his opinion of him to Beverland when the latter visited the Museum in 1692. Beverland took it upon himself to humiliate Woodward 'about a week ago … in a Booksellers Shop … before 4 or 5 Gentlemen who all of them knew me', by repeating to them Lhwyd's alleged words that

one Woodward at London was setting up for a Collector, but that he cryed down all men, & run down every thing … I was, I must confess, somwhat surprised at the relation: 'twas a Caracter I was not overfond of, & I think I do not deserve; especially from you … I am of a temper free & unreserved, & tis possible … I might call a Spade a Spade, & if that was the foundation of your Character, 'twas groundless, & a breach of the Lawes of Sociable Converse, & perhaps of discretion too.[9]

Lhwyd responded to this 'very angry & Expostulatory Letter' by stating that he had praised Woodward as

> a person extraordinary ingenious ... who had ye most considerable collection of English fossils that I had any where seen, & all of his gathering in their native places ... To this character I added that you seemed a litle fond of your own thoughts, but that was a slender fault & would quickly wear of; which I told him not maliciously, or out of Envy to your eccelent parts, but that he might ye readier excuse this fault in case he perceivd it in you ... This I own was an imprudent slip \(& perhaps a mistake)/ but not worth his telling you, or you expostulateing with me so severely ... However in regard you are of a very hot & passionat temper, & of a conversation to me somewhat disagreeable: & yt I also have a greater share than does me good, of that Haste ye common Proverb bestows on my Countrey men; I think our best course will be \to/ let fall our Correspondence. (G, pp. 168–9)

Since each had much to gain by it, Lhwyd and Woodward had resumed their correspondence by the end of 1692, Woodward telling Lhwyd in December that he had discovered Beverland's true character: 'he is neither worth your notice nor mine.'[10] Each was now wary of the other; as Lhwyd later told Lister, 'Softening our pens by degrees we became at last reconcil'd, but how firm Friends I know not, nor am I very solicitous about it' (G, p. 253). Woodward's lengthy letters remained friendly in tone and full of offers and helpful advice, though Lhwyd would scarcely have appreciated Woodward's unsolicited 'hints and cautions towards your conduct' in revising Camden, particularly his assertion that the British, 'a very simple, & Rude people', could not have constructed Stonehenge.[11] In his next letter Woodward reiterated that classical writers showed the Britons to be 'miserably wild & savage people that ... painted their Bodyes, in like manner as the wild-Indians now do ... these People thus rude and savage & destitute of the most common conveniencies of life ... had no Temples, no Houses to dwell in, liveing in Woods & Groves.'[12] Woodward's airy dismissal of Lhwyd's close friend Aubrey, 'for aught I know a very worthy Person; but ... I may freely pronounce him mistaken, if he contend that Stonehenge was a British or Druids Temple', was characteristically tactless. Lhwyd continued to keep Woodward at arm's length while expressing his scorn to trusted friends; in February 1694 he told Lister 'I receiv'd not long since two letters from him, so full of Pride and self-conceit, yt I find he rather improves than corrects ye fault I observ'd at our first acquaintance; and his correspondence begins \now/ to be tedious' (G, p. 218). In June 1694 Woodward outlined to Lhwyd his forthcoming 'Discourse' which would

> begin with the fossil shells, which I prove to be real, & shew by what means they became bureyed in the Earth & lodged in stone &c: wherein I differ from Dr Lister and Dr Plot, for both of whom I have a very great respect ... I treat them with all imaginable honour, & medle not with their opinions any farther than just I am necessitated.[13]

Lhwyd's personal antipathy did not prevent him from acknowledging Woodward's expertise in the 1695 *Britannia,* where he said that he would not provide a detailed discussion of fossils since he was 'expecting shortly a particular Treatise of the origin of form'd Stones and other Fossils, from an ingenious person, who for some years has been very diligent in collecting the Minerals of England, and (as far as I am capable of judging) no less successful in his Discoveries'.[14]

At the beginning of September, Woodward expressed his hope that Lhwyd would look through his book before it went to the press.[15] Lhwyd asked for Ray's advice, which was that he should 'accept the perusall of Mr Woodward's papers, lest he should interpret your refusall to proceed from some sinister principle or disposition in you towards him & his undertaking ... & friendly to advise him of what defects or mistakes you discover therein'.[16] Lhwyd chose to ignore this advice, telling Lister at the beginning of October that he had 'exhorted [Woodward] to put it in the presse with all Speed', Lhwyd's reason (or pretext) being that such perusal would cause further delay.[17] In mid-February 1695 Woodward told Lhwyd that his book was 'almost printed off … in a fortnight they will be done, & some Exemplars bound, & then you may be sure of one with the first of my friends',[18] and on 5 March told him that the book 'is now done' and asked him to send his thoughts on it 'very freely'.[19]

Lhwyd's own catalogue of fossils progressed more slowly than Woodward's work. By April 1691, about a year after first meeting Woodward, Lhwyd had mentioned to Ray and Lister the idea of 'a *Lithologia Oxon.* or a small tract of such figured stones as may be found within 20 miles of Oxf. whether Oxf.shire or Berkshire & Buckinghamshire' (G, p. 139). Ray urged Lhwyd to broaden its scope:

> I would not have you confine your self to so narrow a compasse as ye neighbourhood of Oxford. but take in all of your knowledge that are found in England. I know of no man so fit for such an undertaking as your self. You must promise a generall discourse about ye Originall of those stones.[20]

Lhwyd sent Lister in mid-June more details of the *Lithologia*, 'a Methodical Enumeration \& Description/ of such stones as I could discover within 20 or 30 miles of Oxford … considering first their matter *ex. gr.* Free stone, flint, Peble, Selenite, fluor, Siderites &c. & then their figures'. By broadening its scope so that it became a *Lithologia Britannica*, it might encourage 'setting curious men at Work abt these sort of Enqiries in several Parts of the Kingdome'. Preparing such a work would require 'two Summers travailing at least; which (had I a purse to bear it) I should willingly undertake, leaving a Deputy at ye Musæum, & allowing him twise as much as ever Dr Plot allowd me, towit half ye Perquisites'. Lhwyd would need

> to find out 10 or a douzen Gentlemen who for a hundred species of form'd stones, may be willing to contribute five pounds a peice. They may have two or more of \most/ Spec[ies]. & some stones so rare as perhaps no men in Europe can shew them besides. fifty sorts they may have next Febr. or March upon the receipt of

50 shillins, & the rest on the Michaelmasse following upon receipt of ye remaining part of the money. (G, pp. 145–6)

Lhwyd did not pursue this plan but continued to collect fossils, organise his specimens and read the relevant literature. In August 1691 he asked Lister to 'favour me in yr next with a Catalogue of such authors (especially such as you may suppose unknown to me) as have treated of stones. It matters not in what language; for I have some skill in high & Low Dutch, French, Sp. & Ital.' (G, pp. 148–9).

To make the *Lithologia* as complete as possible, Lhwyd sought the cooperation of collectors in other parts of the country. In early December 1691 he asked Richardson, 'If your neighbourhood affords any formd stones, I beg that you would please to communicate some of them for I have been perswaded by Doctor Lister & Mr Ray; to put into some order what Observations I have made in that kinde' (G, p. 151). At the end of the month he told Lister that Woodward had promised 'all the assistance he can affoard; and perhaps Dr Plot will communicate something of his Kentish collection' (G, p. 152). Lhwyd, meanwhile, had been reconsidering the nature and purpose of his *Lithologia*. Botanists could identify their finds in works such as Ray's *Synopsis*, but there were no such field guides for fossil hunters, nor any accepted system of nomenclature. Fossils were identified by their appearance and assumed function, or by their popular, often folkloric, names. Lhwyd could compare his discoveries with named specimens at the Museum but other collectors lacked a guide to enable them to recognise their finds, ideally in the field. At the end of December 1691 he told Lister that his catalogue would be 'a methodical Enumeration & Description of all manner of figured stones ... of use for Lithoscopists to carry with them into stonepits, Gravelpits &c'. This would be reflected in its format, a portable octavo volume of 'some such bulk as Rays *Synopsis*' (G, p. 152). It would be given 'some such title as *Prodromus Lithologiæ Britannicæ*' in homage to the *Prodromus* of Nicolaus Steno (1538–86), published in Latin in 1669 and in English in 1671.[21] Steno's 'preliminary dissertation' established the modern discipline of stratigraphy and also expressed the view, shared by Hooke and Ray, that fossils were the remains of living organisms.[22]

Since the project would entail 'enqiries in several parts of the king-dome', Lhwyd hoped to train workers to search on his behalf. The problem, as he told Lister in December 1692, was paying them:

> I have made two or three countrey Fellow's here excellent Lithoscopists ... They are labourers in stonepits & are fit to make any discoveries in this kind; tho they can not as much as give their \new found/ stones, any English nick names. One of them who is a shoemaker, would migrate from one Town to an other in this countrey, & sometimes work at his trade & as ye weather servs, search Quarry's, Gravelpits &c. if we could make him a contri-bution of seven shillings a week. But we can make up but three shillings at Oxford, betw. two of us. viz. Mr Archer[23] \of Queens Coll./ and my self. (G, p. 172)

By February 1693 Lhwyd and Archer had enlisted three more contribu-tors, in Wales, Cumberland (Nicolson) and Yorkshire (Richardson). The only 'pensioner' they regularly employed was the above-mentioned itinerant shoemaker, John Smith (or Smyth). Lhwyd's condescending references to him as 'our Witney stone-gatherer' and 'our country-philosopher' show that the relationship was that of master and servant: 'I dispatch'd the Witney merchant to Grantham &c. this day 7 night, and allow'd him five weeks to be absent and no more' (G, p. 193). When Lhwyd lost contact with Smith in the spring of 1693, he callously observed that 'Should he be lost in ye late great snow, or otherwise, his wife & children must (I doubt) go a begging' (G, p. 177). Smith responded intelligently to Lhwyd's instructions: 'He is willing to goe to any part of the Kingdome; & how fit he is to be employ'd you'l be able to Judge from what he has allready performed', Lhwyd told Lister in April 1693 (G, p. 184). Indeed, Smith's knowledge and business sense led to difficulties; Lhwyd complained to Lister in April 1694 that 'our Witney Lithoscopist betray'd us, in not onely instructing Mr Morton,[24] how to find out *Siliquastra*, fish-teeth and Vertebræ; but also in sell-ing to him several curious stones, which of right belong'd to us' (G, pp. 233–4). The final breach came in the summer of 1695. In mid-June

Lhwyd told Lister that he had heard 'that Dr Woodward employs the Stonepicker we sent into Lincolnshire; indeed I have scarce seen him since, and his wife refuses to tel me where he lives'.[25] By late August Lhwyd had 'found him at a Village call'd Sherbourn about 2 miles beyond Warwic.' He had 'got a good Collection, but would not part with them at any reasonable rate, so I have dismissed him ... These Countrey Fellows I find will sell their best Customers for a halfpenny' (G, p. 280). Smith's trade continued to flourish after this, since Lhwyd told Petiver in 1704 that 'you may for a Crown have enough [fossil shells and *Lapides Judaici*] ... from one John Smith of Witney.'[26] The only other stone collector employed by Lhwyd who can be identified is Edward Cozens, the cook of Jesus College. It is not known when he was first employed, but in May 1698 he told Lhwyd that he had followed Bobart's instructions

> to go to Magdalen clay pit and likewise to Cowley a Lithoscoping an send you what I could ... the weather being wet and snow falling almost every daye for a weeke the ground whas cold and I could not lye upon it but I have done my endeavour, I have noe body now to goe a long with me anywhere which makes the way home seeme a little dull.

Cozens had received some education since he referred to the Woodward controversy discussed below: 'I was not long agoe at the booksellers shop and happen to read a sort of scurilous phamplet against LP, wherein he exstoles Dr Woodward but if I had him in my clay pit I would ... confund him.'[27] Other workers were employed on a casual basis; during 1695 Lhwyd had 'severall people' gathering 'stones' at Faringdon.[28] As quarrymen came to realise that Lhwyd would pay for specimens, they sometimes offered to sell him their finds. In June 1695, Lhwyd told Lister that 'a stonedigger shewd me lately a curious vast *Nautilus* dug up 5 miles hence. I bid him half a crown for it', but before he could buy it, the fossil was snapped up by another collector (G, p. 276).

Despite Lhwyd's aversion to theorising, 'I belive a classical enumeration of what Form'd Stones my collection consists of, and a few figures of the rarest, will be more to the purpose, than a Rabbinical chimera about the thawing of the Globe &c.' (G, p. 299), he found he could not avoid discussing the nature and origins of fossils. There were two contemporary views; either fossils were abiotic, *lapides sui generis*, or they were organic, the petrified remains of creatures and plants. Since *lapides sui generis* could in theory be generated anywhere, the abiotic view avoided the difficult questions raised by the view that fossils were the remains of living creatures, such as why marine fossils were found on high mountains or within rocks in the depths of the earth, or why so many fossils could not be matched with extant creatures or plants. This last question raised the dangerous issue of extinction. If an omniscient and omnipotent deity had created the world, creation must then be complete; in such a perfect world, species could not become extinct or be newly created. On the other hand, if fossils had been and were still being formed in the earth, it would be appropriate to speculate about their purpose and how they could have been generated by natural processes. Scientists today would not consider discussing the 'purpose' of fossils to be a valid (or indeed sensible) query. Even in the seventeenth century, the forced, if not desperate, responses to the query should have suggested that the question was inappropriate. These responses included the claim that fossils were the Creator's initial failed attempts (but if omniscient and omnipotent, how could He have miscalculated?), or that these striking imitations of real objects had been created, rather like flowers, 'to beautify the World with these Varieties' for the Creator's pleasure and the delight of humankind.[29] The second question, concerning the formation and distribution of fossils, could be investigated by natural historians by seeking analogous phenomena in the natural world. Thus discussion of fossil origins became part of a broader debate about spontaneous generation,[30] which included the generation of fossils in the earth, of stones in animal and human organs, and, despite the experiments published by Francesco Redi (1626–97) in 1668, of insects. Just as Hooke had attempted to petrify wood,[31] Lhwyd experimented in 1695 and 1701 by burying shells in the Museum yard and elsewhere to see how they might grow or change,[32] reasonable experiments since the six thousand years allowed by the Mosaic chronology implied that if fossils grew, their growth had to be rapid. In his *Oxford-Shire*, Plot had maintained that fossils were stones created in the earth by some 'extraordinary plastic virtue, latent in the Earth or Quarries', the crystallisations of mineral salts which fortuitously resembled natural objects. Lhwyd had initially espoused this explanation, telling David Lloyd in January 1686:

> Naturalists contend much about ye original of these stones; ffor most of them affirme they were once Shells, & therefore call them petrified shells … ffor my part I am soe farre of the contrary opinion yt I think it all most an absurdity to beleive they ever were shells, not doubting but that they are lapides sui generis yt owe their forms to certain Salts whose property 'tis to shoot into such figures as these Shell stones represent. (G, pp. 79–80)

Within a few years, his reading and observations compelled him to acknowledge that both interpretations raised intractable issues, misgivings which, as shown in Chapter 4, he revealed to Ray in November 1690. After reading Ray's views on fossil origins in his *Miscellaneous Discourses concerning the Dissolution and Changes of the World* (1692), Lhwyd changed his mind: 'As to the fossil Oysters … they do, I must confess, confirm me in my apostacy; for I have been inclined to a misbelief of their being mineral forms, ever since I found the first *Ichthyospondylus*,[33] viz. above a year since' (G, p. 171). Ray's book and his own observations also led him to question accepted views on the age of the earth.[34] He accepted Ray's account of denudation by 'rains continually washing and carrying down earth from the mountains',[35] since he had seen how even the hardest rocks in Snowdonia were 'wasted' by 'the rains and snow' which fell there 'in ten times the quantity they do on the lower hills and valleys'. After describing the 'vast stones' on the valley floors he pointed out that since 'but two or three' of the thousands of rocks in Llanberis and Nant Ffrancon 'had fallen in the memory of any man now living, in the ordinary course of nature we shall be compelled to allow the rest many thousands of years more than the age of the world' (G, pp. 157–9). Unlike Halley,[36] Lhwyd did not choose to develop an argument for a greatly expanded time-span but believed that the remains of prehistoric forests – 'I have often observ'd them my self at a low ebb, in the Sands betwixt *Borth* and *Aber Dyvy*'[37] – and the distribution of marine fossils could not be explained by a universal deluge; only an extended period of time could have brought about such changes.

Lhwyd always made a clear distinction between theory and 'conjecture' or speculation. A theory of fossil origins based on examination of specimens could be offered, but to proceed from that to a more comprehensive cosmogony would be to indulge in unwarranted speculation. In later seventeenth-century England, discussion of fossils became entangled with a separate (but related) issue, the need to challenge the naturalistic account of the Earth's development governed by fundamental laws of nature as set out by Descartes in his *Principia Philosophiæ* (1644).[38] The first such attempt to reconcile such an account with Scripture was made by Thomas Burnet (*c.*1635–1715) in his *Telluris theoria sacra*. The Latin text of the first part, dealing with events up to the present, post-diluvial Earth, appeared in 1681, a slightly abridged English translation being published in 1684. Burnet's influential book (which did not mention fossils) was widely criticised by some for emphasising the Deluge, rather than the Fall, as the key Scriptural event, while to others Burnet's emphasis on the workings of the laws of nature made him little better than an atheist. It also raised inconvenient questions, such as the precise amount of water needed to drown the Earth, a problem raised by Beaumont in *Considerations on Dr Burnet's Theory of the Earth* (1693), the manuscript of which he had shown to Lhwyd in April 1691.[39] Lhwyd told Ray in February 1692, 'I must confess I cannot (as yet) reconcile [Burnet's] opinions either to Scripture or reason' (G, p. 159), and in October referred dismissively to 'Dr Burnet's hypothesis, or some such other invention' (G, p. 167).

Woodward's book *An Essay toward a Natural History of the Earth*, published in early 1695, set out a cosmological theory which, unlike Burnet, explained the origin of fossils. It claimed that in the Deluge all solid things, apart from fossils which were the remains of pre-diluvial organisms, were dissolved as the law of gravity was temporarily suspended and everything floated in the liquid mixture. When gravity was restored everything sank in layers to form the present Earth, the heaviest layers and objects sinking first and ending up lowest, the lightest finishing nearest the surface. Although the book sold well, Woodward's theory was unacceptable to many scholars, his high-handed presentation making it even more repugnant. In early April, Ray told Lhwyd:

As for Dr Woodward's *Hypothesis*, if he had modestly propounded it as a plausible conjecture, it might have passed for such; but to goe about so magisterially to impose it upon our belief, is too arrogant & usurping. I cannot but wonder to find such a strain of confidence & presumption running through his whole book: & that he should be so highly conceited of an Hypothesis for which he hath no other proof but a negative one, I mean, that those bodies [fossils] must by this means be thus lodged & disposed, because they could not possibly be \so/ otherwise.[40]

Ray also cast doubt on Woodward's scriptural credentials since he had ignored 'the best Expositors' in claiming that the Deluge began in Spring rather than in the Autumn. According to Ray, Tancred Robinson had been even more dismissive, saying that Woodward wrote 'with a high hand, unbecoming his station & character', and accused him of plagiarising the writings of Steno and of Ray himself. Woodward's 'notion of Gravity' was ridiculous, and Robinson wondered

how his shels should sink lower than metals in ye great Fluid, or how the whole Fossil part of ye Globe should be dissolved in ye Deluge when as the Animal & Vegetable remained entire & untouched. *But the revealing of these secrets he res[erves] for his greater Work, to set us longing for the publication of it.*

He added

I take Mr Lhwyd of Oxford [to be a] man of another temper, & greater both skill & discretion, & wonder his friends do not presse him to publish his Lithology, wherein I expect to find matter of fact and soundnesse of Judgement without any Chimerical whim-sies or Castles in the Air.

Nicolson also expressed his reservations to Lhwyd:

'Tis wonderful that all solids ... should be dissolv'd in the Universal Deluge; and yet the several Species of Shells still keep their primi-tive figure ... There are forty more choak-pears in the Book; which I can by no means gett swallow'd. But – I have done with this Author; being resolv'd never to subscribe for his larger Volumes.[41]

Having remarked that Woodward's book was 'common' in Cambridge bookshops, Archer told Lhwyd 'My opinion of it is, 'twould be of no great damage to real knowledg if, with Dr Burnet's Theory, it were thrown into the Abyss.'[42] After complaining about Woodward's 'publi-clie villifying Mr Wray', Lister was similarly dismissive:

I read so farre of the booke 'till I came at the world, being dis-solved into a mudd, the shells excepted; which wild fancie so shockt me that I desid[ed] to have to doe noe more neither with the authour, nor his writings. These bold stroakes have been in all ages the bain of natural Philosophie: and they will prevail ... for ever, with the idle & prating part of man kind.[43]

The first response to Woodward's *Essay* appeared as it was being published. Knowing that the *Essay* was in the press, William Wotton (1666–1727) published a lengthy abstract of *La Vana Speculazione* (1670) by Augustino Scilla (1629–1700) in the January–February 1696 num-ber of the *Philosophical Transactions*. This 'innocently' laid the ground for subsequent accusations by others that Woodward had plagiarised Steno (already familiar to an English readership) and the hitherto little-known Scilla. The first direct attack on Woodward's book came in *Two Essays Sent in a Letter from Oxford, to a Nobleman in London* by 'L. P. Master of Arts', which had appeared by June 1695. The criticisms of Woodward's *Essay* by 'L.P.' were only part of a thoroughly radical work which combined Spinoza's dismissal of the authority of the Bible as a source of inspired information with the pre-adamite hypothesis of La Peyrère (1596–1676).[44] Oxford gossip suggested it was the work of the Deist John Toland (1670–1722),[45] while in London it was believed to be the joint effort of Lister and Plot (L and P), or of Tancred Robinson, who denied he was the author while admitting that he had assisted L.P. 'upon a Private Disgust' at Woodward.[46] Lhwyd did not know the identity of 'L.P.', telling Lister 'I am sure by some words I find \in/ it, you can guesse at the Author' and adding that he was certain nobody at Oxford could have written it.[47] Even so, Lhwyd was indirectly involved in *Two Essays* since pages 6 and 7 reproduced a passage from a letter of his to Ray which Ray had innocently passed on to Robinson. Woodward wrongly believed that Lhwyd had sent the letter directly to 'L.P.', and may well have been stung by the description of the Diluvial flux as 'the *Hodg-Podg*, or *Pudding* of the Deluge'.[48] Henceforth Woodward regarded Lhwyd as a deadly enemy, and Lhwyd himself, though feigning injured innocence, regarded Woodward in a similar light.

'L.P.' raised several specific objections, insisting, in the name of Bacon's new science and 'rational religion', that the Mosaic descriptions in Genesis of the Creation and the Deluge were not to be understood literally. They belonged to the world of fable and romance, not to the field of objective studies: "'tis no wonder to see *Physical Theorists*, and *Hypothetical Speculators*, grope and stumble in the dark, as soon as they begin to desert the day-light of Sense, and to float out of all depth.'[49] L.P. did not believe fossils to be the traces of living creatures deposited by the Deluge which, according to Woodward, melted the earth 'sometime betwixt the 17th of *May*, and the end of the same Month':[50] rather, they originated from some 'Plastick Power ... of the Earth'.[51] 'S.G.A', the bishop and naturalist St George Ashe (1658–1718), criticised the *Two Essays* in two notices in *Miscellaneous Letters, giving an Account of the Works of the Learned* in December 1695 and February 1696. He rebuked 'L.P.' for a nonsensical treatment of the subject, obvious to anyone 'who is half a philosopher, to which Title L. P. ... can hardly pretend, for he most unphilosophically lays down that Shells, Plants, Insects and other Seminal Creatures may sometimes and in some places be formed and produced without Seminal Parents'. The battle intensified when Thomas Robinson (d. 1719) published *New Observations on the Natural History of this World of Matter, and this World of Life* (1696). Dedicated to Nicolson, who had sent Robinson the works of Burnet and Woodward, this accused Woodward of taking most of his material from the works of Steno and Scilla. After pointing out Woodward's errors and plagiarisms, Robinson pointedly added:

> Much also may be perform'd by the Learned Mr *Edward Lhwyd* ... who hath been very diligent and accurate in his Observations on these Bodies [fossils], and whose Candor and Modesty, joyned with his exquisite Judgment, render him capable of such an Undertaking. (Sig a8b)

The accusation of plagiarism was developed at length by Wotton in 1697 in the second part of a book published jointly with John Arbuthnot (1677–1735), *An Examination of Dr. Woodward's Account of the Deluge*. Arbuthnot's critique was, perhaps, the most judicious response to Woodward's book, acknowledging its virtues but rejecting the 'mechanics' of the theory. Wotton impaled Woodward on the horns of a dilemma: either he had known and plagiarised Scilla's work, or he had not known it and should therefore be grateful to have it brought to his attention.

Woodward's critics were answered by John Harris (*c.*1666–1719) in *Remarks on Some Late Papers relating to the Universal Deluge, and to the Natural History of the Earth* (1697), a substantial 270-page book. Harris easily undermined 'L.P.' by exposing his lack of knowledge of the field, and Robinson was roughly handled. Harris emphasised that Woodward, unlike 'L.P.', confirmed the Mosaic account of the composition of the earth; or, as Lhwyd remarked sarcastically, 'The Dr gives us a far larger Acct of ye state of ye Antediluvian world yn Moses did.'[52] Harris had persuaded himself that 'L.P.' and 'S.G.A.' were one and the same because of stylistic similarities, which he called 'LPisms', in the work of S.G.A. Although very critical of Lhwyd's 'arguments against the Reality of the Fossil Plants', he did not attack him personally.[53] Since it was known that Woodward had assisted Harris in revising his account of the theory, Woodward's enemies could claim that he was the real author of the book.[54] Thus in the margin of page 1 of his copy of the *Remarks*,[55] in response to Harris's accusation that the letters 'L.P.' were not the true initials of the 'lurking' author of the *Two Essays*, Lhwyd noted: 'D.W. guilty of the same thing taking I.H's name' (translated). Since Lhwyd touched only lightly upon his response to Woodward's *Natural History of the Earth* in the marginalia on his copy of Harris's *Remarks*, they do not indicate how deeply disturbed he was by Woodward's book. While on the *Archæologia* tour, he noted in his interleaved copy of Woodward and elsewhere a range of objections to the work and stylistic 'proofs' that Harris and Woodward were one and the same.[56] Lhwyd may have been planning to publish a pamphlet satirising Harris and Woodward, but Tancred Robinson attempted to persuade him to avoid personal attacks and produce a reasoned refutation of Woodward:

> Every body here is convinced that D. W. is the Author of Harris his Book, at least the greatest part of it; therefore I think you need not descend to so many particular comparisons to prove it, least the Reader should grow weary with such similitudes … That which I long for most is your own observations and annotations … If they will make 3 or 4 sheets without all the particular comparisons it may be enough.[57]

By March 1698 Lhwyd had agreed to avoid personal attacks, Robinson telling him 'I am of your opinion that t'will be best to neglect D.W. and his Scaramouch H[arris]', but if he had his 'comparisons and Notes upon them at hand' they might provide Arbuthnot and Wotton with 'some hints and informations'. He also suggested that Lhwyd should not take much notice of Woodward in the preliminary remarks to his catalogue apart from correcting his errors in 'fact or observation'.[58]

His obsession with Woodward and his book led Lhwyd to waste much time and energy on a detailed hostile analysis reminiscent of his father's denunciations of 'Oswestry Malignants', noted in Chapter 2. This was the consequence of suffering for years a fellow researcher whom he was unable to like or respect; and, in fairness to Lhwyd, it should be remembered that Woodward's arrogance made him many enemies. In 1697 Lister became so enraged that he 'drew upon Dr Woodward in Westminster Hall and had it not been for another Dr in company with them (who suddenly interposed) there had been Philosophical blood spilt.'[59] In June 1719 Woodward and Dr Richard Mead (1673–1754) fought a duel after disputing about the best treatment for smallpox.[60] Accounts vary, but it appears Woodward, the first to draw his sword, was slightly injured. Characteristically, he claimed he had enjoyed the upper hand until his foot slipped.[61]

Lhwyd had expected to have his own book ready for the press by March 1695, but further interruptions delayed its publication until 1699. Lhwyd's belief that additional specimens might clarify matters meant that collecting continued well into 1695. He told Lister in August 1695 that he had 'lately taken a Ramble with Mr Morton [clearly forgiven for suborning Lhwyd's lithoscopist] as far as the Humber', and 'brought away a considerable collection of fossil shels, some vertebræ, &c.' Hugh Jones (1671–1702), 'my Servitor and Deputy at the Museum', was sent to search chalk pits in Kent for 'curious stones' such as those Woodward had sent Morton, and to Sheppey to search for *Glossopetrae* (G, p. 280). As an interloper, Jones was roughly treated by the copperas gatherers of Sheppey, Lhwyd telling Lister in September 1695 that 'He return'd a week sooner than I expected, but his reception at Sheppey was such yt I can not much blame him' (G, p. 283).

Another reason for the delay in publishing his book was that Lhwyd wished to discover more specimens of what appeared to be plant fossils from what we would now categorise as Carboniferous-period rocks. Robert Wynn II of Bodysgallen had shown him in (or before) 1690 'a kind of thin \friable/ slat out of ye Cole-mines of Denbighshire, which had \more/ elegant draughts of the branches of some plants, yn any artist can give us; tho I must confesse the plants they resembled (if there be any such in nature) are unknown to me' (G, p. 106). Visiting coal-pits at Acton in Gloucestershire with the Danes in 1691 he found plenty of 'fern branches in ye Cole slat … yt had been dug at least 24 yards deep … also Hartstongue, a Kinde of Cinquefoyl, some other

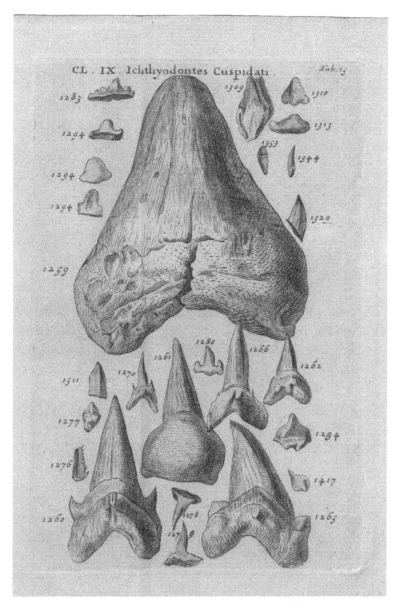

FIGURE 3 *Glossopetrae*, fossil teeth of sharks. Those
from Sheppey were collected by Hugh Jones

capillarys [ferns] & some other impressions yt at present I Know not what to compare to' (G, pp. 138–9). These 'mock plants' or 'lithophyta' were impressions and thus different from the 'variety of Stone yt have ye perfect grain of Wood' that he had previously found and believed were 'wood petrified' rather than *lapides sui generis* (G, p. 161). Lhwyd was intrigued by 'subterraneous plants', and in May 1693 he suggested to Lister that Richard Waller (*c*.1660–1715), editor of the *Philosophical Transactions*, who was 'very curious in designing', might

> setle a correspondence at Acton in Glocestershire, and other colepits, to have all possible variety of mock-plants transmitted to him. Which is a thing perfectly new; and perhaps as unaccountable and as pleasing as any phenomenon in Nature. A *Hortus Subterraneus* would be a Surpriseing Novelty, to ye other parts of Europe. (G, p. 193)

Following his rapid tour of south Wales in August–September 1693, discussed in the next chapter, after complaining about the difficulty of extracting fossils from 'hard limestone or bastard marble', Lhwyd told Lister that

> The colepits of Glamorganshire, affoard as much variety of subterraneous plants as those of Gloc. & Somersetsh. &c. It's observable yt those plants are generally Capillaries. The workmen indeed told me of Ash leavs with ye keys to them; tho perhaps that would prove also some florid fern. (G, p. 204)

Emery suggests that since Lhwyd could not take specimens back with him to Oxford, Archdeacon John Williams (1649–1701)[62] arranged for him to be sent at least two consignments of fossils from the Eaglesbush Colliery, near Neath, during the winter of 1693–4.[63] At the end of July 1694 Lhwyd told John Lloyd that he had seen 'impressions of distinguishable species of plants on cole slates at 20 fathoms depth', twice as deep as those he had seen at Acton. Images of three of the 'Mock Plants', including part of a *Sigillaria*, were included in the engraving of

'Curiosities' at the beginning of the Welsh section of *Britannia* (reproduced in G, p. 241). They were also illustrated in plates 4 and 5 of the
Ichnographia, and were described in 'Lithophyta', pp. 11–14. Miners had
long been familiar with fossil plants; Richardson told Lhwyd in April
1697 that as he was 'buisy in serching ... for these curiositys ... there
came to my assistance *Tho: Kirkman* a skilfull colier ... [who] tould me
that he had taken notice of these sort of bodys nigh 20 years agoe & had
shoune them to several persons', adding that 'in the very place where
I was diging he had taken up a stone with the impresion of a *Braken*
upon it (as he called it) wherein were very discernable both root stalke
& leafe.'[64] Lhwyd replied that 'coal plants have been observed by the
workmen long since, tho' they escaped the notice of naturalists who
till this last century contented themselves with bare reading and scrib-
ling paper.' He added that 'Mr Williams, Archdeacon of Cardigan ...
told me [in May 1694] he had observed much finer patterns 25 years
since in the coal pits of Glamorganshire than some that I shewed him'
(G, p. 335). Welsh miners called such stones '*Carreg Redynog*, i.e. the
Ferny Stone', a term that has apparently not survived. Lhwyd told Ray
in September 1695, when sending him a box of stones from the Forest
of Dean, that the impressions of leaves were difficult to preserve since
they crumbled away as the wet shale dried out; he had even tried burying 'a great clod' of them in a moist place but 'at the taking of it up, [it]
crumbled to pieces' (G, pp. 283–4). As well as including what may have
been the first illustrations of plant fossils, the *Ichnographia* also included
on plate 4 images of four 'insects'. Although the images are crude, two
of them clearly depict arachnids (G, p. 408).

The main reason for the slow progress in completing his book was
that Lhwyd had agreed in April 1693 to participate in the revision of
Camden's *Britannia*, and as soon as that had appeared Lhwyd had begun
preparations for the *Archæologia Britannica* project. Lhwyd's friends were
concerned by the delay, Ray telling him in February 1696 that Robinson
had urged the immediate publication of his 'usefull Synopsis of figured
stones \& Fossill shels/ which he hath ready by him for the Presse' as a
'reall specimen of his Abilities in naturall History'. The delay enabled
Lhwyd's enemies, notably 'his mortall Enemy' Woodward, to 'give

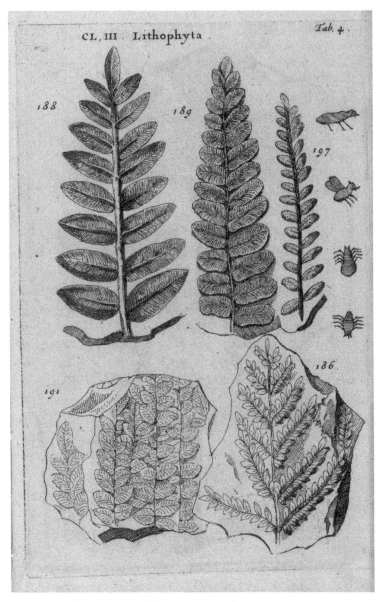

FIGURE 4 Fossil plants, insects and arachnids. No. 188 is from the Eaglesbush Colliery, Neath; No. 191 from Leeswood, Flintshire

out that he neither will nor can perform any thing to the purpose'.[65] Towards the end of 1696 Lhwyd spent a month with his books and specimens in the villages of Stonesfield and Marcham near Oxford, since he found 'the Museum too public a place to mind any thing diligently' (G, p. 320). He had made such good progress that before the end of March 1697 he could send a draft to Lister (to whom it would be dedicated) and to Robinson and Ray for their approbation: "Twill be an 8vo of about 300 pages and will contain 22 copper plates', he told John Lloyd towards the end of March (G, p. 325). Two hurdles remained: finding a publisher and completing the new sections of the book.

Lhwyd had expected the university press to print and publish the *Ichnographia*, an unrealistic hope considering how few 'Delegate's Books', books printed at the cost of the University, it published. The vast majority of titles were 'Author's Books', which were financed by their authors, by subscription (increasingly) or by booksellers.[66] When requesting in February 1698 that the University publish the book he drew the attention of the Vice-Chancellor, Dr John Meare (1659–1709), to its significance: 'in regards it contains the grounds of a new Science in Natural History', it would attract 'a favourable reception amongst ye curious in these studies as well as in foreign countreys as at home' (G, p. 354). Although Lister sent a testimonial to the Vice-Chancellor and the Delegates, and Lhwyd canvassed for their support, the Delegates refused to undertake the work since the press had recently suffered several catastrophically expensive publishing failures,[67] notably Morison's *Plantarum Historiae*.[68] The University was also engaged in a dispute with the Stationers' Company, which had earlier agreed to take five hundred copies of any book printed by the University. Since the Company claimed to have made a loss on every such book recently published, it was not inclined to venture further.[69] In June 1698 the University Printer, Llangadog-born Lewis Thomas, advised Lhwyd to consider publishing by subscription, as was being done increasingly for books 'so much out of the Common Road',[70] and by early July Lhwyd told Lister that he had accepted that the University would not publish his book, nor would 'the London Booksellers medle with it'. Lhwyd, as ever, ascribed the lack of support to the personal failings of others:

> Our present V-C. of Oxfd is ye most pusillanimous I have known
> and \seems/ not much acquainted \with/ any sort of learning out
> of his own Profession: in so much that he never proposd the print-
> ing my Book to ye rest of ye Delegats nor communicated your
> [Lister's] Testimonium ... but \onely/ sent me his advice to print
> it by Subscriptions. (G, p. 377)

Lhwyd considered publishing at his own cost or diverting to the
Ichnographia the subscriptions received for the *Archæologia* project. It
would be expensive: Michael Burghers had quoted a price of 18 shillings
for engraving each plate, a sum Lhwyd considered 'most unreasonably
dear' (G, p. 324). Although more than a hundred copies would have to
be printed to cover printing costs, Lhwyd believed that two or three
hundred copies would sell over time. Members of scientific societies
would want copies, as would booksellers in Leipzig or Frankfurt, and
it could be sold at the Museum, where visiting scholars often asked for
a catalogue of the collection. As an extra inducement, 'we can furnish
ye Buyers with a small Collection of Stones together with ye Book'.
Lhwyd had already given duplicates to friends, fellow collectors and
benefactors of the Museum and had also sold collections of duplicates
to finance his fossil hunting. He explained to Lister:

> I have a private Store House of [stones] besides those which will
> be exposd to view so that whereas they see only two or three of
> a sort: I have in that closet (of many kinds) 50 or 200. for stil as
> I spent my time in making new Discoveries ye old ones would
> dayly occurre which (on such a Design as this is) I always pre-
> servd: and have sometimes sold Forreigners some small collec-
> tions for Guineas a piece. All these matters may be manag'd by
> the Sublibrarian who is an undergraduat, and will not think so
> easy a trade below him; so that it bring him, as well my self some
> Income. (G, pp. 377–8)

He returned to this in the Preface to the *Ichnographia*, where he offered
to sell or exchange duplicates and, in 1708, an advertisement was placed

in the *Philosophical Transactions* listing fossils which could be purchased from Alban Thomas (1686–1771), the Museum's Librarian.[71]

Since Lhwyd had now embarked upon his *Archæologia* travels, his close friends, Lister and Robinson in particular, realised that he could not manage the accounts of subscribers. The idea of inviting subscriptions for the *Ichnographia* was scrapped in favour of a revised plan drawn up in June or July 1698 by influential friends. Sir Hans Sloane told Arthur Charlett, Master of University College, on 4 July 1698:

> Mr. Pepys and I drank yr health this day after & before a great deale of serious discourse on the project about getting the mastery of those who would (and will if not speedily prevented) ruine most good books that are proposed to be printed ... Wee have in ¼ of an hour order'd Mr Floid's book of form'd stones to be printed here wt many cuts. That is to say on the first proposall ten of us subscribed to take of each ten copies at the first cost, provided only 120 were printed 20 of which are designed for the author. (G, p. 23)

Whatever role Sloane and Pepys may have played, the prime mover was Lister, who did not share Lhwyd's hopes of sales and insisted that no more than a hundred and twenty copies should be printed. Robinson disagreed, believing (rightly) that two hundred could be sold.[72] The ten original 'subscribers', all taking ten copies, were Lord [John] Sommers, the Lord Chancellor, Lord Dorset, Charles Montagu, First Lord of the Treasury, Lister, Francis Aston, Sir Hans Sloane, Robinson, 'mounsieur Frouillet of the French Academy',[73] 'Mr Montague's Brother, [and] a German secretary'.[74] Lhwyd himself would receive twenty copies and the plates. When the book was published there had been changes to the list of 'subscribers', Isaac Newton appearing before Lister, and 'Geoffrey, Parisiensis'[75] replacing 'M. Frouilet'. 'Mr Montague's Brother' and the 'German Secretary' had disappeared, leaving nine named sponsors. Perhaps the missing tenth name was that of Pepys. No one believed that every sponsor would claim his full quota of copies, and there was an agreement that Lhwyd could purchase unwanted copies at cost.

The printer was Benjamin Motte (d. 1710), in Ray's view the only London printer who could be trusted with a Latin book,[76] since the other printers there lacked the necessary language skills. Motte, however, was notoriously slow and careless; his dilatory approach had so tried Ray's patience when printing the *Synopsis* that Ray transferred the work to another.[77] Robinson therefore attempted to keep Motte up to the mark by keeping in constant touch with him. Lister was responsible for the plates: by late September most of them had been engraved but none had yet been printed, since the numerals linking the illustrations to the text could not be inserted before the text had been set.[78] On 24 November Robinson told Lhwyd that although Lister's marriage had 'retarded the plates this month', they were all 'carried to the rowling press this morning'.[79] Walter Thomas (d. 1704), Lhwyd's agent for the *Archæologia* project, was jointly responsible with Robinson for the finances of the project and for binding and distributing copies of the book. Despite Robinson's efforts, work did not proceed smoothly. He complained to Lhwyd in mid-October that 'the Printer doth not send me one sheet in 3 to correct, and when I do the workmen will not alter a 4th part … so lazy and ungovernable are they.'[80] In fact, the practical difficulties of editing and overseeing the printing of such a complex volume from afar were far greater than had been anticipated. An additional problem was that since Lhwyd continued to add to the text, he had not been able to send Motte the whole of his book before embarking on the great tour. All the *Epistolae*, closely argued essays comprising a third of the book, were composed during his *Archæologia* tour in 1697. The last part to be written was the Preface, dated Montgomery 1 November 1698. Much of the book had been printed by late November 1698, for when Walter Thomas asked Robinson on 22 November whether a few more copies could be produced (probably to placate subscribers to the *Archæologia*), he was told this could not be done since the type had been distributed.[81] Because of this, it was not possible to correct errors in the text of the book and a list of errata needed to be prepared for printing as part of the last sheet. There was some delay sending Lhwyd ten printed sheets for his corrections,[82] and further delay as his letter listing errata was held up in the post.[83] Thus the *Ichnographia* was published, as Lhwyd complained to Richardson, 'without correcting any of the Errata'.[84] Lhwyd was so dissatisfied with Motte's work that he immediately set about correcting the misprints and preparing material for a projected second edition, entering his corrections and additions in his own copy of the book.[85] His hopes were not realised, and it was not until 1760 that a revised second edition appeared.

Towards the end of December Robinson told Lhwyd how the sponsors had disposed of their copies. Since the text was in Latin, the language of European scholarship, many copies were sent to the Continent. Ten went to Paris; Sloane as Secretary sent copies 'as presents' to Italy and Germany, and Lister presented half his copies to 'Forreigners'. Robinson hoped to obtain the unsold copies, some twenty or thirty in number, for Lhwyd.[86] A letter from Walter Thomas to Lhwyd 2 May 1700 notes the recipients of Lhwyd's copies;[87] some went to influential patrons, others to friends such as Archer and Morton, Lhwyd telling the latter that 'I could scarce secure a Few for my best Friends.'[88] Soon after Rivinus's copy, unbound, was sent to him in Leipzig, a probably pirated edition of the *Ichnographia* was published by [Johann Ludwig] Kleitsch & [Moritz] Weidman, a well-known Leipzig press, its title page bearing the date of the original edition, 1699. Although a very inferior production with particularly poor engravings, this edition indicates the European demand for Lhwyd's work. In Britain, the book quickly sold out, Richardson telling Lhwyd mid-May 1699 that a friend had sent to London for a copy 'but ... they were all disposed of'.[89] It subsequently became very scarce, a correspondent telling Richardson in 1731: 'I cannot by any means, nor for any price, get Lloyd's *Lith Britannicum*' (G, p. 234). The book established Lhwyd's European reputation and, with his contributions to the *Philosophical Transactions*, prompted Sloane to acclaim him 'ye best Naturalist now in Europe'.[90]

The title page of the book described its aim and contents: *Lithophylacii Britannici Ichnographia*, a classified inventory of the British figured stones and minerals displayed in the Museum's cabinets, together with the provenance of each specimen.[91] Perhaps the most important (and generally overlooked) part of the title page was the quotation

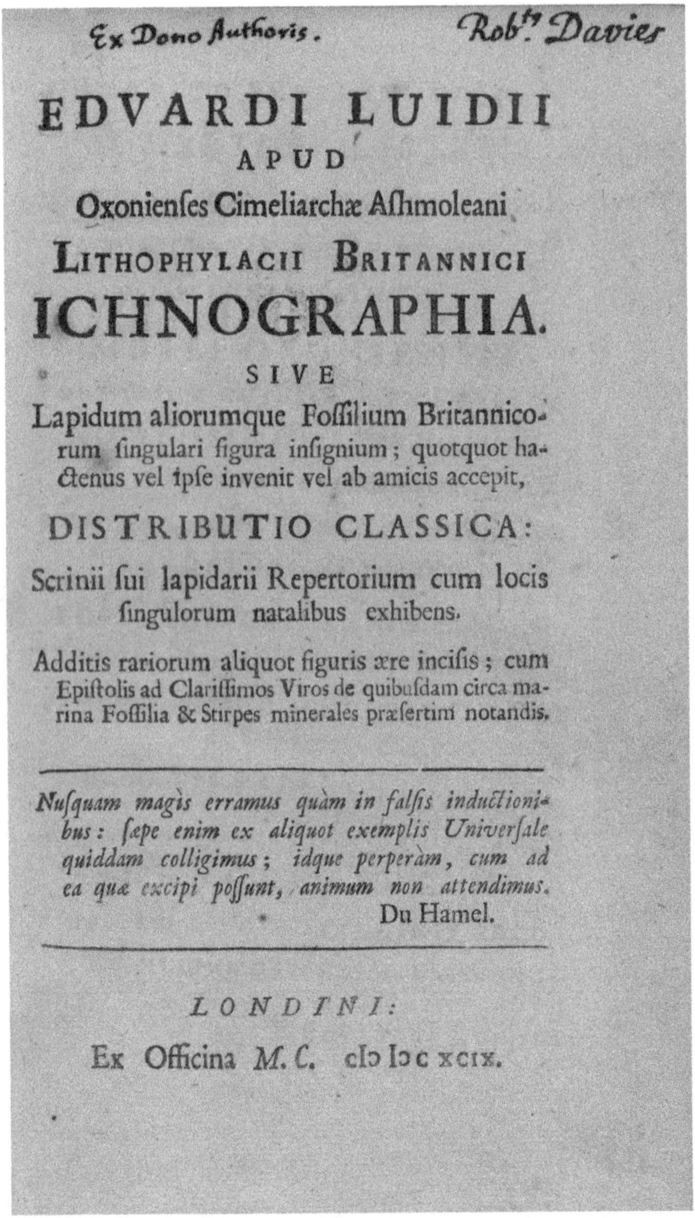

Ex Dono Authoris.   *Rob.ᵗ Davies*

EDVARDI LUIDII

APUD

Oxonienfes Cimeliarchæ Afhmoleani

LITHOPHYLACII BRITANNICI

ICHNOGRAPHIA.

SIVE

Lapidum aliorumque Foffilium Britannico-
rum fingulari figura infignium; quotquot ha-
ctenus vel ipfe invenit vel ab amicis accepit,

DISTRIBUTIO CLASSICA:

Scrinii fui lapidarii Repertorium cum locis
fingulorum natalibus exhibens.

Additis rariorum aliquot figuris ære incifis; cum
Epiftolis ad Clariffimos Viros de quibufdam circa marina Foffilia & Stirpes minerales præfertim notandis.

---

*Nufquam magis erramus quàm in falfis inductioni-
bus: fæpe enim ex aliquot exemplis Univerfale
quiddam colligimus; idque perperàm, cum ad
ea quæ excipi poffunt, animum non attendimus.*
           Du Hamel.

LONDINI:

Ex Officina *M. C.* cɪɔ ɪɔ c xcɪx.

FIGURE 5 Title page of the first edition of the *Ichnographia*,
a copy presented by Lhwyd to Robert Davies of Llannerch

from du Hamel, a warning against false inductions, intended to indicate that the *Ichnographia* was free from such errors.[92] Gunther considered the book, the first comprehensive record of British fossils and their locations, to be an epoch-making work which provided a pattern for later scientific catalogues (G, p. 334). It was also truly innovative as an attempt to open up the Museum collections to all by providing virtual access to its specimens. The 1766 fossils were placed in thirteen classes, each class being described briefly in an introductory section, before the fossils, individually numbered, were described and their locations noted, reproducing their labels in the Museum. Illustrations of almost all the species follow, their numbers keyed to those of the catalogue. Since the collection was already classified, Lhwyd was able to work his way through the collection drawer by drawer, as is suggested in his reference to his assistant's preparation of the fair copy for the printer, 'There are three drawers to be added which my Amanuensis did not copy; but you will easily suppose them from those ten classes that are sent' (G, p. 324). The catalogue progressed from crystalline stones, corallines, and various types of shells, to starfish and sea urchins and to fish fossils. The final category, *anomala sive incertae classis* (anomalous or of uncertain class), contained those specimens, mainly belemnites, that Lhwyd could not place elsewhere.

In the Preface, Lhwyd established his scholarly credentials by naming those who had encouraged him in his studies, Lister and Plot in particular, noting that acknowledged experts had scrutinised the work and emphasising his own experience of collecting over a period of fourteen years. He stressed the need for more researchers since, as he said in *Epistola* III,

> it is impossible that those Sciences should make any rapid advances, when there are but few engaged in the study of them ...
> Tis very rarely we are able to penetrate into subterranean caverns; nor in such secret recesses is it an easy matter to find what are the most worthy our observation. There is therefore great Occasion for the labours of many, & those too in different Countries: nor is it in the power of one man alone fully to settle the mere rudiments

of this science; or the work of one age to render this as compleat as other parts of Natural History. (G, p. 364)

Although Lhwyd stated in the Preface that the book was designed to attract and assist newcomers, his main aim was to further what he had characterised in 1698 as 'the new Science in Natural History', by serving as a record of finds and as a contribution to the debate about the nature of fossils. The most striking feature of the work, as Hellyer emphasised, is the total disjunction between the first, descriptive, section and the theorising of the second, which presents Lhwyd's thoughts on fossils and their origins.[93] These are set out in five essays written as letters (*Epistolae*) to friends: to Rivinus, 26 March 1698; Nicolson, 20 April 1698; John Archer, 1 May 1698; Robinson, 15 July 1698; Ray, 29 July 1698, together with a final letter from Richardson to Lhwyd, 16 June 1698.[94] It is not clear whether Lhwyd's letters were originally composed in Latin, but contemporary English versions or summaries of some of them exist.[95]

In *Epistola* I, Lhwyd reported on his recent discoveries, taking particular pride in the figure of what appeared to be the skeleton of 'a flat fish', found near Dinefwr Castle, Llandeilo, Carmarthenshire. He had already described this to John Lloyd in February 1698, where, after complaining that the 'Country people' might have intercepted Lhwyd's letters to him because they suspected he and his assistants were 'employed by the Parliament in order to some farther Taxes, & in some places for Jacobit spies', he described finding:

> several new sorts of Figured Fossils; amongst which ye enclosed \Figure of some Flat Fish/ represents one of the greatest Rarities hitherto observ'd by ye Curious in such Enquiries. We found plenty of them (thô few fayr Specimens) in a Stone pit near Mr. Gr[iffith] Rice's (wm you remember at ye College) in Caermarthenshire. (G, p. 356)

In early June 1698 Lhwyd had sent Lister a number of illustrations of crinoids from Caldey and pentacrinites from the banks of the Severn

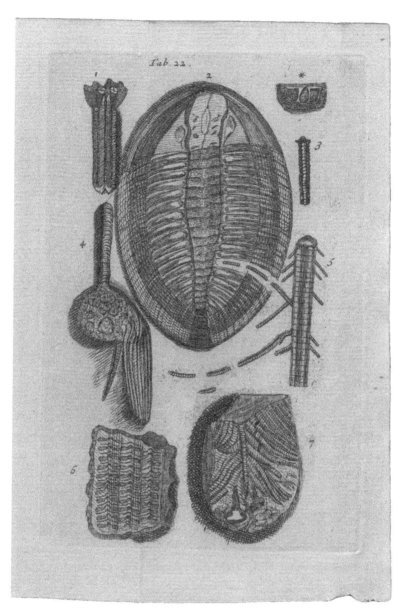

FIGURE 6 Trilobite found at Dinefwr

in Gloucestershire and Goldcliff in Monmouthshire. Two drawings were of incomplete trilobite head-shields, which so baffled Lhwyd that they were depicted upside down. Pride of place went to a find from Llandeilo: 'The 15th whereof we found great plenty must doubtless be referred to the sceleton of some flat-fish', a reasonable supposition since he knew of no comparable living creature (G, pp. 398–400). What Lhwyd had believed was the endoskeleton of a fish was, in fact, the exoskeleton of an extinct arthropod. These were the first descriptions and drawings of trilobites from Wales, the drawing being accurate enough for the specimen to be identified and described by Alexandre Brongniart in the 1820s as what is known today (after many changes in nomenclature) as *Ogygiocarella debuchii*.[96] Writing to Lister in August 1698, Lhwyd commented on the illustrations:

> The rest are *Modioli* or *Vertebrae* of Sea starrs; for I have been long since fully satisfied that all sorts of *Entrochi* & *Asteriae* must be referrd thither; not that I conclude that either of these, or any other marino-terrestrial Bodies, were ever realy either Parts or *Exuviae* [cast off skins or shells] of Animals; but that they bear the same Relation to the Sea-stars, that *Glossopetrae* doe to the Teeth of Sharks; the Fossil Shells to the marine ones. (G, p. 399)

He developed and expanded this view in *Epistolae* II and III, where he analysed in minute detail the structure of all the Echinodermata (crinoids, echinoids and asteriae) that he had examined, and compared them with published examples. Although many specimens were fragmentary, he concluded that the basic similarity in their structure demonstrated that starfish were the origin of 'star stones'. In *Epistola* II he maintained that new fossils 'served much to confirm and establish my Hypothesis … that all the *Entrochi* of whatever species were originally parts of Star fish':

> Comparing this [*Encrinus Lachmundi*] with a Fragment of one of \the/ Rays \of a/ Star-fish; I find it bears the same Resemblance

to it, as the *Glossopetræ* & *Bufonitæ* do to the Teeth of Fishes; the *Cornu Hammonis* to the *Nautilus*; and other sea-productions found on Land do to the real Shells, Teeth, & Bones of Animals. (G, p. 359)

In *Epistola* III he reiterated his view that many fossils were of organic origin:

> by searching for Materials requisite for this work, we trace with astonishment the sea in the midst of Mountains, & discover deep in the Earth whole woods condemned to the mines. For we not only see daily new sorts of shell-Fish, & bones of Marine Animals, together with Trees and leaves of plants, but we also meet with many Fossil Bodies which for many ages were rankt with Minerals, \that/ upon a more nice survey, sufficiently prove their origin to have been Animal, or even vegetable. (G, pp. 364–5)

He hoped that Archer would accept his opinion 'that all Star-stones derived their origin from Star-Fishes' but confessed to uncertainty concerning Belemnites (G, p. 367).

Epistola IV responded to Robinson's queries about marine fossils, and described how plant fossils were found not only in coal and its overlying strata but also in other rocks (G, p. 379). The essay over which Lhwyd took most pains was *Epistola* V to Ray, where he outlined the debate on fossil origins and the alleged effects of the Deluge. He began by expressing objections to the two received views, setting out ten objections to the opinion that the Deluge had deposited fossils where they are now found. Most of these objections had been made by earlier naturalists: that the Flood would have left stones and shells on the surface of the earth not deep within it, and on the floors of caves, not their sides and roofs; that marine fossils were commonly found high up on hills and mountains, whence the rains of the Deluge would have washed them down. Petrified traces of land animals were not being discovered, but if 'all perishd in the Deluge, the Spoils of ye Land might be expected (in Proportion) as well as those of the Sea' (G,

p. 383). Lhwyd was intrigued by 'the vast Number of unknown Marine Fossils so commonly met with throughout most Counties of England; such as we have nothing like, neither on our Sea Shores, nor rak'd by dredges out of the Bottom of the Sea' (G, p. 386). He examined similar difficulties in relation to 'Subterraneous Leavs' by asking how leaves, which one might have expected to float in an inundation, are 'found at ye Depth of at least 20 or 30 Foot'. Lhwyd then examined the alternative view, that 'all these Bodies are form'd in the Earth'. He rapidly dismissed an idea that he himself had espoused a few years previously: 'the great Difficulty it labours under, is that we find our selves incapable of giving any satisfactory Account of the Causes and Manner of such a Production.' Plot's theory of the 'Plastic Power of Salts' could not explain the vast variety of fossil forms (G, p. 389).

Lhwyd then proposed, tentatively, 'some Conjectures I have of late Years entertain'd concerning the Causes, Origine, & Use of these surprising phenomena'. He wondered whether some figured stones might derive from 'fish-spawn, receivd into the Chincks and other Meatus's [openings] of the Earth in the Water of the Deluge', and whether 'the Exhalations which are raisd out of the Sea, and falling down in Rains, Fogs, &c.' might carry with them the 'Seminium or Spawn of Marine Animals'. Since most of the fossil plants are the 'Leaves of Ferns and other Capillaries, & of Mosses & such like Plants', their minute 'seeds', which were 'not at all to be distinguish'd by the naked Eye', could similarly permeate the crevices of rocks and give rise to 'mineral plants' (G, p. 387). He developed his conjecture in five arguments before raising ten objections, to each of which he produced a counter-argument. Lhwyd presented his view as fairly as he could, but his own obvious uncertainties were not likely to inspire others with confidence in his theory. Despite his own doubts, Robinson reassured him in June 1698: 'You need not boggle at your Hypothesis, which tho it will not hold yet may prove a better then any yet propos'd, especially in your modest and doubtful manner.'[97]

Lhwyd was unnecessarily apprehensive for another reason. Sometime in 1696, Richard Mostyn of Penbedw[98] seems to have drawn his attention to some of the theories of Sir Matthew Hale (1609–76), probably in *The Primitive Origination of Mankind* (London, 1677).[99] Replying in December 1696, Lhwyd said:

> I was surprized to find you quote Sr Mathew Hales, for what I thought had never been suspected by any person before. However I am glad to have jump'd in the same opinion with so considerable a person; 'tis enough to extenuat the error whatever absurdity may attend it: but pray acquaint me in your next who told you of this Hypothesis, for I remember not that I writ to you any thing about it. (G, p. 318)

But by 1698, when Lhwyd's hypothesis was about to be published, he began to fear accusations of plagiarism. Robinson reassured him:

> I thank you for your Hypothesis, which I cannot find Judg Hales Orig. in the same manner as you state it; Indeed he says seminal ferments may be convey'd into the Earth by Winds, floods, vapours, rain, and these encrease; but Sir Matthew doth not speak so clearly, or so like a Philosopher as yourself. He knows nothing of the Fissures, etc.

He supported Lhwyd's conjecture by claiming that just as stones grew in human organs,

> the Seeds or Animalcula might as easily augment in the wombs \or bowells/ of the Earth; and seeing your assertion doth not suppose any equivocal or fortuitous Generation, I know not why it might not pass. Some will object how can a Sea Animal live, or grow out of its element, or a shell without a living creature included; I say where the fact appears, reason ceases; Tis a folly to conclude a thing impossible, because it is brought about by unknown means, or by methods inconsistent with the Hypotheses of some men.[100]

Since Lhwyd remained uncertain, Robinson reassured him again almost a year later that although 'Mr Motte has made some gross mistakes, yet in the main every Body commends the Work very highly, except 2 or 3 Mountebanks, who make mouths at your Hypothesis, which still is worth alone all their writings, tho I am afraid it will not hold water.'[101]

Lhwyd's explanation, though unacceptable even to close colleagues, brought together recent claims by microscopists, such as Antonie van Leeuwenhoek (1632–1723), that animalculae could be discerned in rain and human semen, and the new ideas about the hydrologic cycle, recently propounded by Halley, that vapours rising from the sea and carried by winds to mountain tops were precipitated and created springs underground.[102] Lhwyd had long been considering whether rain produced by such vapours could carry seeds which might germinate in some form. When he read the account of insects in amber in P. J. Hartmann's *Succini Prussici ... Historia* (Frankfurt, 1677), he made a note in Welsh on the margins of his copy:[103]

> Is it not possible that the seed of silver fir and pine trees vaporise, and falling in rain and sinking into the earth cause those trees to grow in peat bogs and that these trees produce amber beads according to their nature; just as they 'sweat' resin and pitch above the surface of the earth? (translated)

A letter dated 29 May 1698, probably to Richardson since it discusses 'Insects in the Coalslate', makes it clear that it was Hartmann's account of 'umbrages' or 'mock insects' in amber that had interested Lhwyd.[104] Lhwyd's theory attempted to reconcile two opposing views. He considered that fossils had an organic origin and that, though not all had living counterparts, they were somehow related to the 'real' specimens with which naturalists were familiar. On the other hand, there could be no doubt that they were indeed mineral, and he was not prepared to believe, as did Steno or Hooke, that they were the petrified remains of once living creatures. Although incorrect, Lhwyd's hypothesis drew on recent research and was in tune with his view that observed phenomena had a naturalistic explanation which it was the duty of natural philosophy to seek, rather than to resort to the easier option of attributing them to witchcraft and magic.[105] In whatever field he worked, Lhwyd's conviction never wavered that the explanation of phenomena lay in the close observation and study of as large a collection of classified samples as possible; his guiding principle was that explanations should be rational, not exceptional or supernatural.

Lhwyd disliked dogmatism, for there was always the possibility of new evidence. As he told Ray in *Epistola* V:

> ye frequent observations I have made on such bodies, have hitherto affoarded litle better satisfaction, than repeated occasions of wonder and amazement; for as much as I have often (I may say continually) experienced, that what one day's observations suggested, was by those of ye next calld in question, if not totally contradicted & overthrown. (G, p. 381)

In presenting his theory of formed stones he noted possible objections:

> for they who have no other Aim than the Search of Truth, are no ways concern'd for the honour of their Opinions: And for my part I have been always, being led thereunto by your Example, so much the less the Admirer of *Hypotheses*; as I have been a lover of Natural History. (G, p. 393)

He similarly told Morton in 1700:

> As to the Hypothesis which you say you think I doe not my self belive; my Answer is that I never professd the Beliving it; and that I onely offer it as my Suspicion & doe expressely call it such. I own indeed I have thought it might be true; but never pretended beyond a verisimilitude; and any Body is freely welcome to remove it to make room for Truth which ought to be the onely Scope of all our Observations ... how absurd so ever it may prove ... 'twas my own conjecture grounded on the Observations publishd in that Letter.[106]

BRITANNIA, 1695

When Lhwyd was appointed Keeper in 1691 he was mainly interested in expanding and organising the Museum's collection of 'formed stones' and contributing to the debate on their origin. Although he wished to publish the *Ichnographia* as soon as he could, it was not his only concern, since he was becoming recognised as the expert on the place-names and antiquities of Wales and on the Welsh language. By emphasising antiquarian and philological questions, Lhwyd's work on *Britannia* gave a new direction to his interests and brought him into a closer connection with Oxford scholars of Anglo-Saxon who were investigating similar matters.[1]

Proposals, published in 1692 and revised in April 1693, announced the intention to publish a new and revised English version of William Camden's *Britannia* at £1 6s. 0d., later raised to £1 12s. 0d.; its publishers were to be the London booksellers, Abel Swalle[2] and the brothers Awnsham (1658–1728) and John Churchill (c.1663–c.1714), publishers of several substantial and successful scholarly works. *Britannia*, first published in Latin in 1586, was revised and expanded six times by Camden himself, the last authoritative text being that of 1607. The first English translation by Philémon Holland, incorporating (apparently unauthorised) additions, was published in 1610 and reprinted in 1637.[3] *Britannia* was intended 'to elucidate the topography of Roman Britain, and to present a picture of the Province, with reference to its development through Saxon and medieval times',[4] its framework being the Celtic tribal areas as recorded by classical geographers within which the later English and Welsh shires were grouped. Its Roman basis was emphasised by the increasing number of British and Roman coins and inscriptions added in each revision. *Britannia* soon became the fundamental text for the study of local history, topography, place-names and antiquities, while its stress on fieldwork and unbiased, accurate descriptions of monuments set a standard for subsequent county surveys or 'natural histories'. Like many contemporary antiquaries, the young Lhwyd had transcribed topographical and antiquarian material from Camden, probably in 1680–1.[5] The publishers of the new edition – for this was a commercial venture designed to make them a handsome profit – had realised that the vogue for natural history and antiquarian studies had created a lucrative market for such works.[6] They were therefore prepared to venture the enormous sum of £2,000 on the project, well over £300,000 in present-day value.[7]

Edmund Gibson (1669–1748)[8] was involved in the project from the start, but his initial role (apart from finding potential contributors) is unclear, since the publishers' first choice of editor was James Harrington (1664–1693), a promising young lawyer described by Lhwyd as 'a Gentleman of vast acquaintance and Interest'.[9] Following Harrington's unexpected death in November 1693 the publishers invited Gibson to oversee the work. Realising that he was in a strong position, Gibson negotiated terms that ensured scholarly standards and consistency.[10] Unlike Harrington, who had no real experience of antiquarian studies, Gibson was a prominent member of the group of Anglo-Saxon scholars associated with Queen's College, Oxford, and had established his scholarly credentials in 1692 with his edition of the Anglo-Saxon Chronicle.

Revising *Britannia* was far beyond the capabilities of a single person if new, authoritative material were to be included for each county. Even before becoming editor Gibson had set about gathering a team of scholars, turning first to the Oxford Anglo-Saxon group before casting his net more widely. He assembled a notable team of natural historians and antiquaries including Aubrey, John Evelyn, Nicolson, Thomas Tanner (1674–1735), White Kennett (1660–1728), Ralph Thoresby (1658–1725) and Plot. Gibson believed that the revision's attention to natural history, especially botany, would improve sales: 'Mr Ray's

Catalogue of Local Plants will secure us the Botanists and Natural Philosophers.'[11] Ray provided botanical descriptions for each English county but did not cover the flora of Scotland or Ireland. He initially hoped that Lhwyd would cover Wales,[12] but because of Lhwyd's 'unwillingnesse of being at unnecessary trouble' (G, p. 242), Ray supplied a two-page appendix, 'More rare Plants growing in Wales', mainly plants found in Snowdonia and Anglesey.[13]

Gibson told Lhwyd on 15 April 1693 that 'Mr Swalle, the undertaker of the English Camden, is now in town to procure persons that may carry on that work.' He hoped that Lhwyd, to whom he had been introduced in November 1691,[14] would 'not be wanting in your assistance towards the revising Wales' (G, p. 178). Since Lhwyd was concentrating on completing the *Ichnographia*, he responded coolly to the initial invitation to revise descriptions for part of north Wales but told Lister he thought he might 'pick up some materials \from ye Gentry and Clergy/ which may prove usefull an other time' (G, p. 193). Although he did not explain what he had in mind, he had told John Lloyd in March 1692 that he had 'a strong fancy that I may (if it please God I may live 7 years longer) meet with some encouragement towards ye writeing some part of the *History of Wales*' (G, p. 161). Lhwyd may also have discussed his plans with Plot, for in May 1692 the latter suggested that the Principal and Fellows of Jesus College would support a project on the Natural History of Wales. Plot's tactless comment that 'there are perhaps some particulars relating to that affair wherein you are not as well skilld as is necessary, but if I had you with me but one summer, I should not doubt but to render you as able for such an Undertaking as any man in England',[15] would not have improved Lhwyd's opinion of his former superior. Always short of money, Lhwyd could not afford to ignore the prospect of being paid for his work on *Britannia*, and accordingly told Swalle that 'I would doe something for two or three Counties' (G, p. 193).

Lhwyd's initial intention was to cover Denbighshire, Merioneth and Montgomeryshire, the counties with which he was most familiar and where he had contacts. By the summer of 1693 he learned that 'the Gentlemen yt had once undertaken ye other three counties of N. Wales' had withdrawn (G, p. 208). As he reminded Gibson in September 1694, Lhwyd then 'thought it better, with the Assistance and Advice of Friends, to offer my best Endeavours, than to leave it wholly to the Management of some person lesse acquainted with the Language and Country' (G, p. 244). On 4 August 1693 he told Lister that he had

> offer'd [Swalle] to doe all Wales; & to take a journey speedily quite through it, for ten pounds in hand; and twenty Copies of the Book, when it shall be publishd: but he'll not come up farther than ye one half of it. I have now sent him my last resolution, which if he does not accept I shall break off with him. But if we agree ... I'll begin my journey towards Monmouthshire, on ye 14th. (G, pp. 197–8)

The tour was necessary because Lhwyd lacked contacts in south Wales and had never travelled farther south than Cardiganshire (G, p. 143). He prevailed over the publishers and made a seven-week tour of south Wales during August–September 1693. Neither Lhwyd's working notes nor sufficient letters have survived to allow a detailed reconstruction of his itinerary, but he appears to have travelled through Monmouthshire, Glamorgan, Breconshire and Carmarthenshire to Pembrokeshire and south Cardiganshire. After returning to Oxford Lhwyd remained suspicious of his publishers, telling John Lloyd in October that 'Mr. Swall and Mr. Churchill (who are my taskmasters) did not require I should put myself to ye trouble & expences of a journey into Wales; for they care not how litle is done for that Country; their business being only to procure subscribers, which they have allready done to their satisfaction.' He also complained that correcting Welsh place-names on the maps 'would take up \much/ more time and pains than [the publishers] are willing to requite ... Whatever I can adde or correct otherwise, I'll spare no pains' (G, p. 198).

Thereafter his reading and particularly his correspondence became the main vehicles for his research. Lhwyd's preferred research method had always been fieldwork and personal observations, but he also drew on the knowledge of local informants to supplement his own investigations or provide data where a personal visit was not possible. Ideally they would be sufficiently well read to appreciate the nature of Lhwyd's research but, as he told John Lloyd in July 1695, humbler sources of knowledge also deserved respect: 'As for Caer Verwyn I can not see why an illiterat Shepheard may not be believ'd in such a case as soon as a Bishop; for ye names of mountainous and desert places are better known to those of his profession than men of learning' (G, pp. 278–9). Such a respect for reliable local testimony can be seen in Lhwyd's handling of the mysterious Merionethshire fires of 1693–4. On 5 January 1694, his old college friend Maurice Jones (b. 1663), rector of Dolgellau, sent him a vivid description of

> the unaccountable fiering of a matter of eleven or twelve hay-ricks in the night time most of which were reduced to ashes thô some part of some of them were sav'd, the 1st of them tooke fier upon X'mas eve & the rest within the compass of a weeke afterwards … they all tooke fier of the same side next the sea, and not withstanding that the Hay-ricks themselves were burnt by the violence of the fiers, yet there was no men received any hurt who interpos'd their endeavours to save the hay, notwithstanding that they labour'd hard in the fiers.[16]

Lhwyd replied promptly, but his naturalistic explanation failed to convince Jones, who wrote again a month later providing more detail and explaining why he rejected Lhwyd's view.[17] The possibility of witchcraft increased the panic felt by the community. People believed that they had saved 'several Ricks of Hay and Corn' by employing traditional remedies since 'any great noise, such as the sounding of Horns, the discharge of Guns, &c. did repel or extinguish it'.[18] Lhwyd discussed the problem throughout 1694 with Ray (G, pp. 235–6), Lister (G, pp. 219, 242) and others, and published an abbreviated version of Jones's account in the *Philosophical Transactions* (G, p. 221). Mid-August 1694 Jones sent Lhwyd up-to-date information provided by a local gentleman, William Wynne of Maes-y-neuadd, Llandecwyn, near Harlech,[19] whose son, Robert, also sent Lhwyd an account.[20] Lhwyd combined these accounts when presenting his own theory in *Britannia*.[21]

From the first Lhwyd had been adamant that there had to be a natural explanation. As he received reports of the fires, Lhwyd also heard that locusts had been seen in Wales. He examined one specimen in December 1693 and offered to send it to Lister (G, pp. 211, 215). Locusts had previously been reported in Britain, but Lhwyd, searching for a naturalistic explanation of the fires, wondered whether the fires and the arrival of the locusts were somehow related. He gathered more evidence about the locusts and scoured the literature for analogous occurrences, developing his views in a letter to Lister published in the *Philosophical Transactions*, February 1694 (G, pp. 223–5). He presented his theory, at first 'only a random guess' (G, p. 235), more fully in *Britannia*: 'fiery Exhalations' known as '*Ignis fatuus* [will-o'-the-wisp], *Ignis lambens, Scintillæ volantes*, &c.' were well-known phenomena which were never reported as being toxic or incendiary, nor did they move so regularly as these fires. 'Wherefore seeing the effects are altogether strange and unusual, they who would account for it, must search out some causes no less extraordinary.' Lhwyd was frustrated that he was unable to examine the fires himself but had to rely on 'a bare relation of the matter of fact'. Nevertheless, he suggested that the mephitic atmosphere might be caused by the decaying remains of locusts reported to have arrived in north Wales some two months earlier. They had been discovered on beaches near Aberdaron, and Lhwyd could quote historical instances of 'these Creatures being drown'd in the Sea, and afterward cast ashore, [causing] a Pestilence'. Kindling fires, however, was another matter, and although 'Pliny says of [living locusts], *multa contactu adurunt*, i.e. they burn many things by the touch', Lhwyd left the cause of the fires an open question. Ray, Lister, Maurice Jones and, it seems, most of his friends, rejected the theory; indeed, Lhwyd himself admitted: 'I know there are many things might be objected, and particular the duration of this fire; but men are naturally so fond of their own conjectures, that sometimes they cannot conceal them, though they are not themselves fully satisfy'd.'[22]

Shortly after Lhwyd had asked his old friend John Lloyd to assist him in replying to Nicolson's queries (see Chapter 5), he sent him a further, lost letter requesting his help with *Britannia*. By the end of May 1693, Lloyd had given Lhwyd the names of Welsh antiquaries who could assist him, notably Humphrey Humphreys, bishop of Bangor, Robert Davies, Llannerch, William Williams, Llanforda, who 'has incomparable Manuscripts in his Custody', and Richard Mostyn, Penbedw, a list which indicates how well informed Lloyd was about antiquarianism in north Wales.[23] He promised to contact Richard Mostyn (1658–1735) on Lhwyd's behalf, expecting 'great satisfaction fm him, not onely because of his own great knowledge & Curiosity, but he may help you to his Brother Sr Tho: Mostyn's M.Ss.'[24] Richard Mostyn of Penbedw, Flintshire, an Oxford graduate, was the third son of Sir Roger Mostyn, first baronet of Mostyn Hall.[25] He was not as active a collector of manuscripts as his brother Thomas (1651–92), the second baronet, of Gloddaeth, but possessed a fine library. Before Lloyd could contact him, Richard Mostyn, who had heard of Lhwyd's role in revising *Britannia*, had indicated his willingness to try to facilitate access to Sir Roger Mostyn's collection of Roman coins. After some delay Lhwyd ventured to approach Richard Mostyn, perhaps in October 1693, since Lhwyd's earliest surviving letter from November 1693 suggests they had already been in touch: 'I hope you'l pardon my boldnesse, if I beg of you some contribution towards Flintshire; or any other part of Wales'. He asked Mostyn about various antiquities, including '*Maen y Chwyfan*, which seem so strange a name that I cannot devise what should be the origin of it … a stone chest or coffin full of Urns … call'd *arffodegaed y wrâch*' (the witch's apronful), and, near the abbey of Valle Crucis, the inscription 'CONGEN FILIVS ELISEG'. Lhwyd concluded, 'But I need not give any hints of what would be acceptable … to one that's a farr better judge of it than my self' (G, p. 207–8). Mostyn proved to be a scholarly correspondent with whom Lhwyd could share his discoveries and discuss their significance. A practical antiquarian, who insisted on accuracy in transcribing inscriptions, Mostyn sent Lhwyd copies of Flintshire inscriptions and descriptions of *carneddi* (cairns) and artefacts. He appears to have been readier than Lhwyd to comment on the broader implications of their researches and on the purpose of writing history: thus he praised John Wynne's *History of the Gwydir Family*,

'J don't know any book give's a plainer image of ye age he write's of, which J take to be ye end of history.'[26] None of Lhwyd's other Welsh correspondents reveals the same breadth of outlook, perhaps because Mostyn's circle, which included English antiquarians and scholars such as Hearne, was far wider than that of most of Lhwyd's other Welsh informants.[27] Lhwyd appreciated Mostyn's assistance, telling John Lloyd, 'I take \Mr./ Mostyn (betwixt you and I) to have as good a Share of both of these [candour and judgement] (besides his other Qualifications) as any I have had correspondence with, in Wales' (G, p. 243). Lhwyd's respect for Mostyn's views is obvious in their later correspondence, when Lhwyd had become increasingly interested in questions of language history and relationships. For his part, Mostyn was pleased to support Lhwyd's work both intellectually and financially, but, being a genuinely modest man, insisted that his name should not appear,

> for wt J chiefly value my selfe upon, is living & dying private. J am not of Tully's mind, who thought happiness to be either in *otio cum dignitate, aut in negotio sine periculis* [leisure with dignity, and in business without danger], for my part Jm'e contented with *otium sine indignitate* [leisure without indignity].[28]

Lhwyd respected this request and Mostyn is not mentioned in *Britannia*.

Writing to Mostyn in November 1693, Lhwyd referred to the use of a questionnaire: 'I have sent to some Friends a few general Queries which I hope have come to your hands' (G, p. 208). From about 1650 onwards questionnaires had become a popular method of collecting data, employed by individuals and by learned societies.[29] The Royal Society supported their use, publishing in its *Philosophical Transactions* in 1666 Boyle's influential 'General Heads for a Natural History of a Countrey'. In spite of their limitations, deriving from their framers' preconceptions and from the lack of control over the quality of responses, questionnaires were an invaluable source of information which enabled researchers to draw upon the knowledge of a wide range of people. Plot, Lhwyd's mentor, had already made good use of them in his studies of Oxfordshire and Staffordshire.[30]

It is not clear when Lhwyd sent out his Queries, but a letter dated 28 May from John Lloyd may represent his first response to the document. Here Lloyd proceeded to relate folklore associated with local churches, the first of a number of letters full of topographical lore, place-name legends, inscriptions, and plans and measurements of *caerau* (forts) and circles.[31] The copy sent to David Lloyd was enclosed in a letter dated 1 June,[32] and John Lloyd said on 10 June that he had 'dispersed several copies' of it.[33] An early response, dated 12 August 1693, from Thomas Evans contained a range of information, some of it supplied by the bishop of Bangor.[34] Only one example of the questionnaire is known to have survived, a single folded quarto sheet. Unlike the comprehensive queries used for his later *Archæologia* tour, these were not 'general queries' and, judging by this specimen, each related to a particular county (in this instance Anglesey) and was intended for named individuals. On the recto Lhwyd set out the nature of the project, his own role in it and the scheme's scholarly credentials (this part, no doubt, would have been common to all copies), before requesting on the verso specific items of local information:

Whereas a new Edition of Cambden's Britannia is intended with such Additions & Amendments as are promis'd in the printed Proposals, These are to acquaint such as are desirous of promoting that design in the counties of Denbigh, Merionith, & Montgomery yt Mr Edw: Lloyd, Keeper of the Ashmolean Library and Repository at Oxford, has at the desire of the Undertakers promis'd some Annotations on Mr Cambden's discourses of those three Counties. And therefore it is his request if the Curious wou'd please to communicate such Observations in Antiquities and Natural History as may seem pertinent to the Authors design whether Additions or Illustrations of what he has deliver'd, of Amendment of some mistakes. For since no man can pretend to be free from Error, it will not be any rashness in some cases to question the Authority of this learn'd and Judicious Writer. Whatsoever Remarks shall be communicated towards the compleating of this Work shall be gratefully acknowledg'd, either publicly or privately as occasion shall be offered < > And if any Gentleman residing in the Countey (who for yt and other Reasons may probably be much fitter to perform this task) shall be dispos'd to undertake it, the Person above mention'd upon Notice being given him will be not only ready to desist but also willing to contribute what he may towards the promoteing so generous a design.

But if otherwise the charg of this Province be wholly left to him, he hopes yt all curious Gent: who shall be made acquainted with it, will be willing to contribute their Observations: since the reserving of them (wou'd with some Persons) but lessen the Character our Nation has allways deservd. And because his ability may reasonably be call'd in question; it is to be consider'd that the whole work, tho peform'd by several hands, will be perus'd before it be committed to the Press by one or two of the most eminent Antiquaries of England.

Some General Information that wou'd be Acceptable in this Undertak[i]ng.

An Ennumeration & Description of all the Ancient Towns & Castles in each County, and also any other Old building whether Rom: or British omitted by Camden with a Catalogue of such Places as seem by their Names to have been such thô nothing else at present remain to inform us: and conjecture ab[ou]t ye Origin of the Names of Places.

[Overleaf]

Queries for Anglesey

Whether there be any Grounds for conjecture about ye significa-tion of ye word môn?

Anianthus lapis why call'd in British *maen urael*: < > whether there be any History or Tradition that the ancient Britans had any use of this stone?

Jn what History does it appear yt the Jrish harbour'd in Anglesey upon the Downfall of the Roman Empire? Or whether th[e] assertion of Mr Camden be only conjecture from Tradition and some names of places?

A description of ye Jrishmen's cottages.
Whether there be such a place as ʒn hericy Gwidil (as
Mr Camden calls it) or Hannèrcuis Gwyddel or any thing like it?[35]

Lhwyd's 'Queries' elicited a good number of responses. The quality of the
information and its systematic presentation reveals how well respondents understood the purpose of the queries. Former Oxford students,
particularly those from Jesus College, were particularly qualified to reply
since over the years Lhwyd had inspired several of them to undertake research in natural history on their return to Wales. Thus William
Gambold (1672–1728)[36] sent four letters between December 1693 and
September 1694 describing Castell Nadolig (Cardiganshire), antiquities in Nevern church and village and elsewhere in Pembrokeshire,
St Dogmaels Abbey, and urns and a stone called 'Maen Dewi'. In
October 1695 Gambold apologised that his

> successe in the prosecution of the injunction you were pleas'd to
> lay upon me whilst at Oxon, has been very mean and inconsider-
> able, as having found nothing remarkable but a Periwinkle shell ...
> 4 yards deep in a marl pit ... & also a rude piece of wood.[37]

His copies of inscriptions in Nevern church and St Dogmaels were
included in *Britannia*,[38] and he would provide a comprehensive account
of his parish of Puncheston for Lhwyd's next project.

John Lloyd struck a congenial note in the opening sentence of his
letter of 19 December 1693: 'Without any further praeface, you are
to arm yrself with Patience to engage wth a tedious letter. For J am
resolv'd to omitt nothing tho ne're so trifling, nor to leave off, while any
room is left, nor yet to regard any exactness in penning it.'[39] A detailed
account of his route from Y Bala to Bryn Eglwys in Denbighshire followed, in which he noted ancient monuments and provided plans of
caerau, *carneddi* and stone circles, with their measurements and associated traditions and place-names. Nicholas Roberts (1646–1707) was
an experienced naturalist and antiquary who had maintained a correspondence with Lhwyd since 1687.[40] In May 1690 he had described the

breeding habits of migratory sea birds on the Pembrokeshire coast.[41] He offered to assist Lhwyd mid-August 1693, welcoming the opportunity to revise his earlier descriptions of Pembrokeshire and Carmarthenshire and to provide additional material.[42] He sent a comprehensive account of Pembrokeshire antiquities and natural history to Lhwyd in December 1693,[43] part of which (notably a description of making laver bread) was published in *Britannia* under his name (pp. 57–80).[44] Since Roberts realised that it would not be possible to include such extensive accounts of every county, he may have withheld his account of Carmarthenshire. He complained on 18 December of Lhwyd's 'unkindeness'; although Lhwyd had been within a mile of his house during his south Wales tour, he had not visited him. This was unfortunate, 'For I could have given you a large & better account by discourse in one hour, than by writeing in a quire of paper.'[45]

Robert Humphreys (see Chapter 4) responded in August 1694 to Lhwyd's request for information on local antiquities, providing full accounts of standing stones and cromlechs, and of their traditions, of an inscribed stone and another stone magically guarded by 'some spiteful spirit', of the cave legend of the lost musicians of Llanfair-yng-Nghornwy, the parish where 'the Asbestos is found', and of *Cyttiau gwyddelod* (Huts of the Irish), round stone huts of the Roman or immediate post-Roman period found in north-west Wales. He noted 'natural curiosities', fossils such as the 'Cornua Ammonis' and one similar to 'St Cuthburds bead', and the eggs of 'strange' sea birds. He responded cautiously to Lhwyd's 'judicious, & learned account' of the mysterious fires in Merionethshire, 'wch if it be not the true, is I am satisfied, the most probable yt can be given. I have some weake observations of my own to confirm my belief of the probability of it.'[46]

Maurice Jones commenced his reply to the Queries on 21 August 1693 with deceptive modesty:

> I am heartily sorry that I cannot send you any thing of a satisfactory solution to the Queries you sent about this county. I have lay'd about & enquired as much as ever I could, & after all I can make

neither my self nor any body else much wiser. All the account that I could pick up is as followeth.[47]

There followed four closely written pages about Roman coins, careful descriptions of possible camps, roads and artefacts at Cefn-caer, interpretations of place-names and the route and vestiges of Sarn Helen. 'Beddi gŵyr Ardidwy' and other graves interested him, especially a stone in Trawsfynydd with an inscription that he read as *Hic jacet Euporeus qui homo Xtianus fuit*.[48] He related the onomastic legend of Llyn y Morynion and of 'Ceridwen wrâch' from one of Robert Vaughan's manuscripts. Lhwyd's lost reply apparently sought further clarification while casting doubt on the presumed 'Christianus' reading. In *Britannia*, Lhwyd reproduced his own 1689 reading of the Trawsfynydd inscription, 'HOMO ---- RIANVS FVIT', but recommended further examination since his reading had been made at a time when he was 'wholly unacquainted with any studies or observations in this kind'. Nevertheless, he argued that for a stonecutter to have omitted the letters STI after RI would be 'such a fault as we have scarce any instance of' in published collections of inscriptions (p. 72). He also used the discussion of a lost inscription on one of the stones of Beddau Gwŷr Ardudwy[49] to note some differences between Roman and post-Roman inscriptions, 'I suppose it Roman, For he says it was a polish'd stone ... whereas all the later Inscriptions that I have seen in Wales, are on large Pillars, which are generally rude and unpolish'd' (p. 72). Finally, Lhwyd's account of the Merionethshire fires made considerable use of material supplied by Maurice Jones (pp. 70–3).

Not all responses to the Queries were from experienced naturalists and antiquaries. Some correspondents appear to have sent only one or two letters but they, too, took pains to be as informative as possible. John Davies, rector of Newborough, Anglesey, apologised that since he had little time to consider the queries, Lhwyd should 'take these short remarks as hints to putt you in mind of better conjectures'. His lengthy 'short remarks' discussed interpretations of place-names and gave a plan of the standing stones at 'Bryngwin',[50] based on his visiting the site himself and pacing out the distances. He similarly reported on Bodowyr,

'Another kind of a sepulchrall monument ... still standing on its Tripos soe Tite and pert as if itt had been newly erected and tho we have many of this kind of a far larger extent yet I have seen none soe neat.'[51] Other letters reveal Davies's belief in the druidical associations of many of the monuments he described.[52] His interpretations of place-names, though not dismissed by Lhwyd, are not acceptable to modern scholarship. Davies, however, was reluctant to theorise, preferring to draw inferences from the evidence. For example, that standing stones at a place called Maes Mawr, traditionally the site of a battle, were graves, was proved by the discovery of a helmet there. The druids made human sacrifices but

whence they derived those Rites is not soe obvious, I had once some conceptions in order to the clearing of that point of wch I am not now soe fond in regard that for want of Books I want Authority to back the notion I have of it, & therefore it shall remaine as itt is an abortive Embrio.[53]

Lhwyd used much of this information, sometimes tacitly recast; on the margin of p. 81 he notes 'A Letter from the Reverend *Mr. John Davies*', but this was an amalgam of rewritten letters. Lhwyd's style is more polished than that of Davies, the latter's remark on his 'want of Books' becoming 'but having nothing out of history to confirm my conjecture, I shall not contend much for it'.

John Thomas (1646?–95), rector of Penegoes,[54] covered the whole of Montgomeryshire, describing tumuli throughout the county and 'great ditches & long, between ym in hight not inferior to Clawdd Offa', as well as 'a Circular ditch upon ye top of ye long mountain in Welsh cefn digoll' and stone circles. Venturing as far as the Border, he had examined cairns on Corndon hill.[55]

Archdeacon John Williams regretted that he had not met Lhwyd on his south Wales journey.[56] In November 1693 he described stones

upon a jutting att ye North West of Keven-bryn ... put together by labor enough but no great art into a Pile ... there is a vast unwrought stone (probably abt 20 tun weight) supported by 6 or

7 others yt are not above 4 foot high, & these are set in a Circle some on end; and some edgewise or sidelong.

Noting that the stones were of the local millstone, he speculated how they might have been erected, 'ye pully's & leavers, ye force & skill by wch twas don are not so easily imagind. The Comon people call it Arthur's Stone, by a lift of vulgar imagination attributing to yt Hero an extravagant size & strength.'[57]

Griffith Jones (1666–1749), rector of Llanrwst and schoolmaster there,[58] collaborated with friends and neighbouring clergymen to copy inscriptions as best he could and to interpret place-names. In January 1694 he related local legends about Llywelyn Fawr and included excerpts from a manuscript copy of Sir John Wynn's *History of the Gwydir Family*.[59]

Erasmus Saunders (1670–1724), who would later become a fine scholar, visited Carmarthenshire and his native Pembrokeshire to copy inscriptions and describe antiquities, including, it appears, the Pentre Ifan cromlech, stone circles and standing stones.[60] The precocious undergraduate William Rowlands of Beddgelert[61] provided a comprehensive account, in Latin, of Snowdonia.[62] He later submitted Latin notes on monuments in Beddgelert church, on 'snake stones', on his collection of alpine flora and on a Welsh manuscript grammar by the poet Huw Machno (d. 1637). Lhwyd noted Rowlands as a potential assistant.

Some may have responded to Lhwyd's Queries, despite their inexperience, because they were pressed to do so. In December 1693, Thomas Hancorne (d. 1731), vicar of St Donat's, enclosed his copy of the inscription on the stone bridge at Margam.[63] He had hoped to include remarks on Glamorgan antiquities by 'Rabbi Wilkins', probably Thomas Wilkins of St Mary Church (1625/6–99),[64] but the latter was tardy in keeping his promise, though he had shown Hancorne three manuscripts which he intended to bequeath to Jesus College. Lhwyd replied promptly, sending a further request, to which Hancorne replied in mid-January 1694, enclosing a careful description and drawings of a stone pillar at Llantwit Major church and a copy of an inscription on another which had been scraped and rendered 'almost illegible since you saw it',[65] presumably in 1693. Lhwyd drew upon these descriptions in *Britannia*, adding material on the enigmatic furrow or 'canalicus' in the first stone from a letter sent to John Aubrey by James Garden of Aberdeen.[66]

Since Lhwyd viewed his task as being that of an editor, his retranslation of the Latin of the 1607 text of *Britannia* made few changes. He modernised the style of the English translation and consistently updated Camden's references; thus 'in the memory of our grandfathers', has Lhwyd's marginal note, 'So said, ann. 1603' (p. 24), and 'my present reviewing these notes' is noted in the margin, 'Circ. ann. 1607' (p. 33). Lhwyd was second to none in his admiration of Camden: 'I look upon Mr. Camden to have been one of the most learned, judicious, and ingenious writers in his kind that ever England or perhaps any other countrey has produc'd', adding that 'as to what we can adde or correct ... were he alive ... he would be thankfull for it' (G, p. 201). Although none of the original text was omitted or altered, Lhwyd did not hesitate to correct Camden's errors. The claim (p. 30) that the sun was not visible above the hills at Llanthony until after one is dismissed in a marginal note: 'This is contradicted by such as know the place.' Corrections could involve substantial additions: the 'subterraneous noise' that Giraldus Cambrensis had reported in a cave on Barry Island, an account accepted by Camden and 'many later writers', was dismissed; 'at present there are no such sounds perceived here.' This view was supported by a letter from John Williams, 'a learned and ingenious Gentleman of this Country' (p. 42), who suggested that the noises emanated from the blowhole at Worms Head in Gower.[67] A substantial addition corrected Camden's fanciful etymology of the name 'Emlyn', Llwyd maintaining that 'the name ... was Roman, and is the same as the *Æmilinus* mention'd in Denbighshire' (p. 50). Camden had maintained, on the authority of Giraldus Cambrensis, that beavers had formerly existed in Wales; Lhwyd deployed both literary evidence, 'some old poets', and onomastic evidence, 'Lhyn yr Avangk', to demonstrate that this was so, and that other animals, such as wolves, wild boars and even bears were once to be found in Wales (p. 61). Lhwyd had been pondering linguistic questions since the late 1680s, and now took the opportunity to express his thoughts on orthography and historical phonology, as in discussing the relationship of the river names Taf and Thames (p. 49), the values of *m/v* (pp. 43–6) and the commonality of the Gaulish and British languages (pp. 67, 76), all topics that would come to the fore in the *Archæologia*. He had been given an opportunity to re-examine inscriptions that had deteriorated since he first viewed them and others that he had copied when he was less experienced. He discussed place-name elements like *–wy* in river names, and *-i* and *-o* as indicative of Roman names; for *Bala* (pp. 72–3) he returned to the explanation that he had given in his 1688 glossary (noted in Chapter 4) but with additional examples; he considered the significance and meaning of *llan*, 'enclosure, yard', *carn*, 'a pile of stones', *cromlech*, from *crom*, feminine of *crwm*, 'crooked, bending' (pp. 55, 81), with which he compares names containing *gwyr*, 'crooked', such as *llech* or *maen gwyr, meineu gwyr*. He recognised that this interpretation might cause difficulties where *gwyr* (modern Welsh orthography *gŵyr*), 'crooked', was not applicable to stones standing directly upright or inclining only a little. Since *gwyr* (modern Welsh orthography *gwŷr*) might be the plural of *gŵr*, 'man', he correctly maintained that *Bedheu Gwyr Ardudwy* should be 'the Graves of the men of Ardudwy' (pp. 71–2). Lhwyd never had much patience with popular beliefs, 'fables', except as 'antiquities', nor with unwarranted assertions and opinions, which he rebutted with scientific or historical evidence. He was interested enough in the unusual ox horn, the *Matkorn* (inner part of a horn), that he was shown by the sexton of Llanddewibrefi church, to give a full description of it, but as for the associated legend of *Matkorn yr ŷch bannog* or *Matkorn ŷch Dewi*, 'I shall not trouble the Reader with, as being no news to such as live in Wales, nor material information to others', a characteristically dismissive Lhwydian turn of phrase (p. 60). Tales of sunken cities, drowned in lakes by earthquakes, he regarded as fabulous, 'one of those erroneous traditions of the Vulgar, from which few (if any) Nations are exempted'. Such disasters might have occurred in countries subject to earthquakes and subterraneous fires, 'but since no Histories inform us, that any part of Britain was ever sensible of such Calamities, I see no reason we have to regard these oral traditions' (p. 28). Camden was a severe critic of

Geoffrey of Monmouth, who had inserted into his work 'so many ridiculous Fables of his own invention'. Though Lhwyd accepted that
Geoffrey's *Historia* includes many fables, that these were of Geoffrey's
own invention 'may seem too severe a censure, and scarce a just accusation', since Lhwyd believed Geoffrey was the translator of an ancient,
but still extant, Welsh chronicle, a venerable copy of which could be seen
in the library of Jesus College (p. 31). Lhwyd's defence echoes a view,
held by other contemporary scholars, which was based on a reading of
the colophon of some Welsh versions of the *Historia*,[68] but it also illustrates his pride in his national cultural inheritance. This may be compared with some of his marginal comments in his copy of Humphrey
Lhuyd's *Commentarioli Britannicae Descriptionis Fragmentum*, the most
heartfelt of which are on the inimical nature of some Anglo-Welsh
relationships.[69] Perhaps the most noteworthy feature of Lhwyd's additions is the provision of detailed, objective descriptions of buildings,
monuments and even of stamped Roman bricks.[70] He clearly enjoyed
writing the descriptions of Caerffili castle (p. 40), Maen y Chwyfan
(p. 91), the barrows and tumuli of Carmarthenshire (p. 51), his account
of Merionethshire (p. 68) and the important IDNERT inscription at
Llanddewibrefi (p. 60), to which he would return on a later visit, when
he removed the stone from the wall to read the hidden portions (see
Chapter 8). He refers to 'several ancient Stone-Monuments' in parishes
in north Cardiganshire, 'whereof I shall briefly mention such as I have
seen, because they may differ in some respects from those already
describ'd'. Although he was familiar with the area from his youth, when
he must have seen Gwely/Bedd Taliesin, he nevertheless requested his
kinsman Simon Pryse to make a plan of the monument for him, which
he did in July 1694,[71] too late to be included in *Britannia*. Accurate
drawings of inscriptions and sites were essential. As Lhwyd put it, while
describing the location of a *maen hir* (standing stone) on Cefn Gelli
Gaer, the site 'may be better delineated than describ'd' (p. 41). This was
his usual principle, which explains why so many inscriptions are carefully
reproduced and why he provided a full page of illustrations of various
Welsh 'curiosities' (p. 21). In the *Natural History of Stafford-shire* (1686),
Plot had maintained that the 'new antiquarians' were interested in 'things

not persons and actions'. Lhwyd was similarly prepared to allow monuments and sites, fully and accurately described, 'to speak for themselves'; what they 'said', however, was to be corroborated, as far as possible, by history, not only by chronicles and books but also by material evidence.[72] This kind of interpretation is different from 'conjecture', which lacked any historical corroboration or foundation. Lhwyd thought that since *Kittieu'r Gwydhêlod*, 'vast rude stones laid together in a circular order' were too 'ill-shap'd' to serve as foundations for any higher building, they could not be regarded as evidence for an Irish settlement (p. 80); the name simply reflected the popular tendency to ascribe any unaccountable monument to the Irish, who by tradition 'were possess'd of this Island' before Christianity, just as in Scotland unfamiliar edifices were ascribed to the Picts.[73] After listing half a dozen other place-names containing Gwyddel/Gwyddyl, he concludes, 'having no History to back these names, nothing can be infer'd from them' (p. 80). This appeal to history is a recurrent strain in his comments. His discussion of the barrow 'Krig y Dyrn' at Tre-lech (p. 51) provides an instructive example of his way of working and thinking. Following a comprehensive description of the monument and of the excavations carried out there, he turns to the name, which he conjecturally translates 'King's Barrow', linking the unknown *dyrn* with 'teyrnas' and 'teyrn', 'originally the same with *Tyrannus*, and signify'd King or Prince'. From 'the labour and strength requir'd in erecting it', Lhwyd was 'apt to suspect it the Barrow of some British Prince', since it was 'much too barbarous to be suppos'd Roman' and there was no record of Saxons there, nor of Danes so far inland. That it was pre-Roman is 'only my conjecture', Lhwyd's argument being that, once conquered, the Britons 'had none they could call *Teyrn* or King whose corps or ashes might be reposited here'. Lhwyd combined personal observation, deductive reasoning and book-based research in setting out his arguments step by step for rejecting the belief that *cromlechi* were erected by the Danes as ceremonial sites (pp. 55–6). There was no historical evidence that the Danes ever had any dominion or settlements in Wales; their raids around the coasts would have given them little opportunity or reason to build permanent memorials, and such monuments are also found in the mountains of Caernarfonshire, where 'no History does inform us, nor conjecture suggest, that ever the Danes have been.' Moreover, there are considerable differences between the British examples and those described in the standard reference books of Danish and Swedish antiquities, which contain nothing to be compared 'to that magnificent, tho' barbarous monument, on Salisbury Plain', nor any that have a *kist vaen* (stone chest or coffer); even if there were resemblances in structure, these would not prove that *cromlechi* were Danish, since cultures, both pagan and Christian, had always imitated or borrowed from one another. Lhwyd might offer an interpretation but was never dogmatic: 'For my part, I leave every man to his conjecture' (p. 52), 'what they [the letters of an inscription] signifie, I fear must be left to the Reader's conjecture' (p. 62), 'But I am not satisfied with this notion of it [his first attempt at interpreting an inscription] my self, much less do I expect that others should acquiesce in it' (p. 62), comments that epitomise his open attitude in all aspects of his research. Careful description extended not only to 'curiosities' but to landscape and natural history. Lhwyd's love and deep knowledge of Snowdonia is obvious in his description of the 'British Alps' (p. 74) and his detailed discussion of the gwyniad (*Coregonus pennanti*) of Llyn Tegid (p. 72).

By July 1694 Lhwyd had 'finishd the greatest part of my Annotations ... but am afrayd they will not print all I send [the]m' (G, p. 239), since Gibson had warned contributors in February that he might be compelled to curtail their copy should additional costs raise the price of the book beyond what had been advertised. Because of this, Lhwyd told Mostyn in March that White Kennett, 'who had undertaken Oxfshire, is fallen off, and some others begin to be dissatisfied. Some friends also advise me to break off ... but I'm resolv'd to go through with it as well as I can' (G, pp. 231–2). Although Dr Jonathan Edwards, Principal of Jesus College, had suggested that Lhwyd should 'keep my papers, in order to print them apart', he decided that he would submit his contribution (G, p. 227). When Lhwyd first saw printed sheets of the work, probably of Cornwall and Devon, he admitted that his suspicions had proved groundless, telling John Lloyd on 8 September 1694: 'I am now satisfied we shall have pretty fayr play: for whereas I suspected all this while they would print but few Notes or Additions, I find by some Counties I have seen, their Additions are almost as large as ye Text' (G, p. 243).

Lhwyd wrote up his contributions quickly, county by county, between May 1693 and the end of October 1694 (G, pp. 245–6). By 31 July 1694 he had 'long since' sent the six counties of south Wales and Monmouthshire to Gibson (G, p. 240). Denbighshire, Montgomeryshire, Merioneth and Caernarfon had also been sent by the beginning of August but had somehow gone astray in London. A search by Lister and Gibson of all the London inns used by Oxford carriers and coaches eventually found the missing parcel, which eased Gibson 'of a great deal of trouble and you of noe small Labour'.[74] Despite the tight timetable, Lhwyd had controlled his disparate material, viewing the revision of the Welsh counties as a unified project. Since his accounts of monuments, place-names, inscriptions and other discoveries often included cross-references to examples in other locations or to discussion elsewhere in *Britannia*, what was intended to be a county-by-county inventory of ancient monuments began to assume the character of an integrated national corpus with bibliographical references. Gibson clearly respected Lhwyd's grasp of his material, and the printer's copy (Cardiff MS 4.172) shows that his revisions were of a minor nature. Contributors were not required to read proofs, but Lhwyd was allowed to see the final printed sheets and inserted fifty emendations, mainly correcting place-names. Those of Lhwyd's notes which arrived too late were retained by Gibson and included in the expanded 1722 edition of *Britannia*.[75]

Difficulties arose regarding the number of engraved figures and maps, where costs were an important consideration for the publishers. Lhwyd was happy with the former, telling Lister mid-November that 'Mr Gibson ... has grav'd all the Figures' (G, p. 254). The maps were another question. The proposals had promised a map of each English county, and it had been understood that maps of individual Welsh coun-ties would similarly be included. A map of Monmouthshire, presumably considered an English county by the publishers, was included, but the publishers 'put a trick upon us as to the maps of Wales' by deciding to include only two maps, of north Wales and south Wales. In disgust,

Lhwyd, already unhappy with the orthography of place-names (G, p. 198), refused to correct them[76] (G, p. 259).

Lhwyd's concerns about Welsh orthography had one fortunate result. He explained to Gibson in mid-September 1694 that while he had attempted to follow the sense of Camden's text, 'I have sometimes differ'd in writing the Welsh names of Persons and Places', and outlined his system (G, p. 244). At Gibson's request, the whole of Lhwyd's letter explaining his 'method' for writing Welsh personal and place-names by employing 'a more general Alphabet, whereby such as are unacquainted with that Language, will pronounce the Words much truer; and they that understand it will find no occasion of mistakes', was included in *Britannia*, followed by a guide to the 'Pronunciation of the Welsh'.[77] Lhwyd, who held that language comparison should take account of both meaning and pronunciation, had long been aware that etymologists looking for similarities between words in different languages were often misled by orthographies that concealed their pronunciation. To resolve the problem, Lhwyd drew up, initially for his own use, what was, in effect, a phonetic alphabet, and it is this that was printed for the first time in *Britannia* and later in the *Glossography* (1707), the first volume of the *Archæologia* project.

Britannia was published towards the beginning of March 1695, a second, expanded edition appearing in 1722. Lhwyd's unbound copies of the 1695 edition cost him £1 12s. 0d., bound copies being sold initially for £1 16s. 0d. and then at £1 18s. 0d., the price which Lhwyd asked for his copies. Contemporary responses to the volume largely reflected the bitter and entrenched political divisions of the 1690s. Despite Gibson's studied moderation, nonjurors, notably Hickes, had refused to contribute to a work which they considered to be a glorification of the 1688 Revolution.[78] When *Britannia* appeared it was condemned by nonjurors such as Hearne and Francis Brokesby and by outright Jacobites such as Francis Atterbury. Even they praised the quality of Lhwyd's scholarship, which, Brokesby maintained, Gibson's other contributors had failed to match.[79] Hearne claimed in 1706 that 'excepting what ye Learned Mr Llhuyd ... did, there is nothing of any great moment' in the book.[80] He added in 1713 that *Britannia* was

full of faults both in translation & 'Additions'. Llhuyd's part about
Wales excellent, and as good as anything of Camden's own. This
a necessary supplement because of Camden's want of skill both
in the Irish and British Tongues and his not having travelled in
Wales.[81]

Despite partisan criticisms, the work sold well and was frequently
reprinted. The only criticism Lhwyd himself received was from his
cantankerous friend Nicholas Roberts. He had received his copy by
October 1695 and berated Lhwyd for publishing extracts from his work
since it had been written 'off hand', 'Indeed I am not well satisfied with
your exposeing me.' Some of the Carmarthenshire comments displeased
him even more, since they did not accord with his own observations.
The paper was of poor quality and a couple of maps were missing, making it 'the dearest booke that ever I bought'.[82] Modern assessments of
Britannia emphasise the quality of Lhwyd's contribution, Parry claiming that his 'outstanding' additions 'transform this part of the *Britannia*
from a fairly inadequate sketch ... to the most rewarding part of the
whole volume'.[83]

Revising the Welsh counties for *Britannia* did not represent all
Lhwyd's activities in 1693–4, for completing the *Ichnographia* remained
a major concern. Lhwyd integrated the two projects so that, as shown
in Chapter 6, new discoveries were reported in letters published in the
Philosophical Transactions and also in *Britannia*, where his excitement at
the discovery of 'mock plants' was given its fullest expression. Lhwyd's
revisions to *Britannia* reveal many of the characteristics of his future
work and set him on the way to his most important undertaking.

ARCHÆOLOGIA BRITANNICA: THE 'GREAT TOUR', 1697–1701: WALES

As shown in the previous chapter, by the early 1690s Lhwyd was beginning to think of compiling a 'History of Wales'. The invitation to contribute to *Britannia* in May 1693 reinforced his growing interest in antiquarian and linguistic questions, and the informed assistance he had received in his researches from Welsh gentry and clergy interested in antiquities and natural history showed him that there was a potential market for the work he had in mind. As his plans became known he began to receive offers of support; as he told Lister in mid-April 1694:

> The Principal of Jesus Coll. [Jonathan Edwards (1629–1712)] talks much of a History of Wales, & offers his interest to dispose ye Gentry, to give me ye like encouragement yt Dr Plot met with in Staffordshire. I answear'd yt in two or three years time I hoped to capacitate myself for it, & yt I should then be very glad of such an Employment.

Meanwhile, Lhwyd was learning Irish 'yt I may be ye better Critic in ye British in case I should ever be concernd in ye History of Wales' (G, p. 249, date amended from *EMLO*). The immediate occasion of his study of Irish was the presence in Oxford of John Toland, a native Irish-speaker whom he got to know during the first three months of 1694.[1]

As *Britannia* drew nearer publication, Lhwyd told Lister in July 1694, 'If I gain any credit by this: its not unlikely but our Gentry may be hereafter willing to encourage somthing more considerable' (G, p. 239). He was briefly tempted by the prospect of undertaking paid research in Ireland. Dr Owen Lloyd (1664–1738), the Secretary of the Dublin Philosophical Society, told him in October 1694 that the Society desired 'his correspondence in all Chapters of Natural history'.[2] This prompted Lhwyd to wonder whether he could now undertake the Natural History of Ireland which he had discussed with the Molyneux brothers in 1688: 'When *Camden* is out, it may be guessd partly whether I might be capeable of serving the public in that way ... The onely objection is that such an undertakeing would be too expensive' unless finance could be found (G, p. 256). By January 1695 he admitted to Lister that this was probably wishful thinking, and Lhwyd's hopes were finally dispelled when William Molyneux told him in early February that there was no prospect of 'an incouragement of 60 pounds pr An. in this Place'.[3]

The mirage of Ireland was soon replaced by the prospect of research in Wales. At the end of December 1694 Archdeacon John Williams thanked Lhwyd for accepting his contributions to *Britannia*, and added:

> I often secretly wish yt you had leisure and encouragement to yor own wishes to frame ye naturall History of Wales; and the Civil alsoe as far as it may be retrievd and purifid from ye fabulous traditions of our own Countrymen, or the dry partial accounts of the English writers.[4]

In his reply Lhwyd must have mentioned the need for financial support, since at the end of April 1695 Williams told him that

> the impediments you mention ... which obstruct the designe of a natural history of Wales might ... be surmounted; if ye Welsh gentry were warmly solicited to a standing & Generous contri- bution to encourage & support such a work ... I was informd that Dr Plott had good assistance & reward for his paine att Staffordshire, & I should think such an undertakeing in Wales

would meet with a becomeing recompense. I should be ready to move any likely Gentlemen in this or the neighbering County of Carmarthen very earnestly.[5]

Despite his poor health, Williams employed his local connections to canvass the support of a number of Glamorgan gentry.[6] Their promised support over a period of seven years enabled Lhwyd to plan a wide-ranging history, and they proved to be the most constant of all his subscribers. Lhwyd dedicated the *Glossography*, the first volume to appear, to Sir Thomas Mansel of Margam (*c*.1688–1723) since, as Lhwyd told John Lloyd in 1707, Mansel had been 'so constantly Active in Procuring Subscriptions & Payments, & so much the Author of the Undertaking from the Beginning; that I must have shewd too much Ingratitude if I had dedicated it to anyone else (G, p. 531). Regrettably, although Williams had done so much to make what became the *Archaeologia* possible, Lhwyd broke irreconcilably with him in the spring of 1696 when Williams refused to take his side in his dispute with Woodward.[7]

In May 1695 Lhwyd expressed to John Lloyd his delight with the support he had received from Glamorgan: 'If the like encouragement would be allow'd from each County, I could very willingly spend the Remainder of my days in that Employment; and begin to travail next Spring.' He then outlined his plans:

I shall draw up som Proposals, which may contain a short Account of the Design … I must confesse the Sallary may at first sight, seem too much; and the time of seaven years too long: but such as are acquainted with Natural History, know there's no good to be done in't without repeated observations; and that a Countrey of so large Extent can not be well survey'd, and the Natural Productions of it duely examin'd, under the space of four or five summers; \after which the/ time remaining will be short enough for methodizing the Observations and publishing the History. Besides, during that time I propose to take one journey into Cornwall, and an other into the Scotch Highlands, in order to collect parallel observations;

that so I need not rely much on the credit of forreign Writers ... I
am resolv'd to spare no pains nor charges in the performance; and
therefore unlesse I am enable'd to go through with it, 'twould be
imprudence to medle with it at all. (G, pp. 269–70)

Lhwyd's letter contained the very important sentence 'Nor should
I onely regard the Natural History of the Countrey, but also the
Antiquities.' This change of emphasis was obvious in the proposals
dated 20 October 1695 as *A DESIGN OF A BRITISH DICTIONARY,
HISTORICAL AND GEOGRAPHICAL; With an ESSAY, Entituled,
ARCHÆOLOGIA BRITANNICA: AND A NATURAL HISTORY OF
WALES*.[8] The work would comprise three sections, some four to six
volumes in all. Two dictionaries would be onomasticons, one containing
lists of persons of importance in British history and writers in Welsh,
Cornish or Armorican (Breton), the other of place-names with their
meanings. *Archæologia Britannica* would include a comparison of Welsh
with other European languages, especially Greek, Latin, Irish, Cornish
and Armorican; secondly, a comparison of British customs and traditions with those of other nations; thirdly, an account of British pre-
Roman monuments; fourthly, an account of Roman and later antiquities
with their inscriptions. In other words, Lhwyd intended to push back
the historical horizon beyond the Roman period. Such a study would
have to encompass all the Celtic-speaking peoples of Britain, Ireland
and Brittany, since they represented the 'First Inhabitants' or, to use
Camden's term, the 'ancient Britains'. The Natural History would comprise a geographical description of the country, descriptions of soils,
stones and minerals, formed stones, and native flora and fauna.

Lhwyd emphasised that he could devote all his time to the project
since he was free of family cares, was 'engag'd in no Profession' and was
not '(by the favour of the University) oblig'd to Personal Attendance
in my present station'. He had told Lister some months earlier that the
Visitors would not object to his absence: 'such an Employment either
in Ireland or Wales would not at all oblige me to resign my present
Station: for I am satisfied the Visitors would give me leave on such an
occasion to be absent as much as I pleased' (G, p. 256). Possibly Lhwyd

had in mind the revised Museum statutes which he had drafted in July 1695.[9] These were eventually approved in 1697, but Lhwyd went to great lengths to prevent the Visitors from seeing them in their final form before they were printed, telling his recently appointed Underkeeper, William Williams (1673–1701),[10] in September 1697: 'I would not have you permit the new Vicechancellor nor any of the Visitors have sight of the Paper before Printing for fear of new Scruples ... I am in some doubt whether you had best print it at Mr Halls or with Mr Thomas.' Since the former might 'carry it to the V. Chr.', Lhwyd thought it best to entrust the printing to Lewis Thomas, though he would have to pay him for the work.[11] As a close friend, Thomas told Lhwyd that although printing would cost twenty-five shillings, he would not ask for more than a guinea.[12] The revised statutes were designed to increase the Keeper's income and also make it possible for him to be absent for a long period. Clause 11 read:

> The Keeper for every Month he is Absent from the University, shall present the *Museum* with Ten several Coins or other Antiquities; or else the same Number of distinct Natural Bodies, either different from those in it already, or better Specimens: in Default whereof he shall pay Ten Shillings a Month towards the Library of the *Museum* each Visitation.[13]

Copies of the *Design*, seeking annual financial support for five years for the planned tour, were sent to the gentry and influential people in Wales and London, a few being sent to Cornwall. Woodward did what he could to frustrate Lhwyd's plans. In February 1696 Lhwyd told Lister:

> I find that Woodward is my implacable Enemy, and does me all the mischief he can which yet ... is not much. If he stil maintains his character among the Clergy (for I suppose he never had much amongst Naturalists) he may doe me some prejudice; for I depend somthing on ye Interest of ye Bishops of Bangor and St Asaph. (G, p. 299)

According to Robinson, when Lhwyd's 'Proposalls and Design were recommended to the Royal Society [Woodward] spoke very contemptibly of them there, and I hear he continues the same Insolence in other places.'[14] Following a brief visit from Lhwyd, who gave him a copy of the *Design*, John Byrom (1648–1717) could scarcely contain his enthusiasm, telling Tanner on 30 October 1695:

> I am willing to think yt he will meet wth an encouragemt wch such a work deserves ... Wee shal then, I hope \see/ our 1st planters, as naked as they went, wthout ye imaginary dresse of mere art or fancy: then we shal see old citys & villages rise up from their ruines, wth their ancient names & the Etymologys of 'em. (G, pp. 286–7)

Lhwyd had requested subscriptions at a very inauspicious time as new and increased taxes were levied in the 1690s to meet the cost of the war against France. In 1695–6 the poorly executed Great Recoinage, when old clipped silver coins were called in to be replaced by new milled coins, caused an acute shortage of ready money.[15] As the Wrexham clergyman and schoolmaster John Stodart (1669–1708) told Lhwyd in February 1696, 'there happen'd such confusion amongst our gentry about ye money yt they woud hearken to noe proposals of this kind.'[16]

Even so, Lhwyd was sufficiently encouraged by the subscriptions he had received, about thirty pounds a year, and by promises of support and cooperation, to undertake a preliminary tour of Wales from late April to mid-October 1696. On 23 April Robinson sent Lhwyd a letter posing many queries relating to the natural history of Wales.[17] Lhwyd sent a lengthy reply from Swansea in mid-September. By June, he had reached Bangor, where he met Bishop Humphrey Humphreys and planted roots of rare plants from Snowdonia in his garden. He had found rare maritime and mountain plants and searched in vain for fossils (G, pp. 308–10). His quest for minerals was more successful since he had found a crystal 'about 9 Inches long, and thicker than my Wrist, transparent as Glass for the better half, but opaque about the Root like white Marble'. Smaller crystals were 'of the colour of a Topaze', and he 'was inform'd of others purely Amethystine, found in the Vally of Nant Phrantcon. I find our Ancestors ... made themselves Beads of opaque, or Marble Crystal; for I have one given me, cut like a Lottery-ball, and perforated; found not long since in Meirionydhshire.' He questioned whether the 'transparent stones', called *Ombriæ* by both Plot and Lister, 'are so form'd naturally', and noted one 'lately given to me (set in Copper with a little Handle to it) by the name of *Tlŵs Owen Kyveiliog* i.e. Owen of Kyveiliog's Jewel; so call'd, because found in an old *Crig* or *Barrow*, near the Place where he lived'. Lhwyd later told Humphrey Foulkes that the Tlws had been 'bestowed upon me by Mrs. [Margaret] Pugh of Mathafarn'.[18] The stone 'is the same with the transparent *Ombriæ* in the Museum ... This was formerly called in S. Wales, Glain Kawad' (G, p. 414, date and addressee corrected from *EMLO*). After responding to Robinson's queries about wildlife, Lhwyd said he intended to follow his suggestion to 'try the Barometer and Thermometer on the top of Snowdon and Cader Idris' the following summer. He was forestalled by Halley, who carried a barometer up Snowdon, 'a horrid spot of hills', on 2 May 1697. From the fall in the mercury level Halley calculated that Snowdon's height was 1015 m, 70 m short of the modern value.[19] Finally, Lhwyd drew an important conclusion from his observations: 'What seemed to me most strange, were vast confus'd stones, and ... fragments of Rocks standing on the surface of the Earth, not only in wide Plains, but on the Summits also of the highest Mountains.' These disproved Woodward's theory, since they demonstrated that 'the Mineral Kingdom was never dissolved ... when the matter of Stone was fluid, how it could possibly create solitary Rocks ... without a Mould to cast it into'.

Robinson told Lhwyd that his 'rich cargo from Swanzey ... is sufficient for a volume according to the measure and proportion of some late writers'. He recognised the significance of Lhwyd's negative findings:

I take your not meeting with any figur'd stones in the Welsh Mountains to be a very considerable Observation; for it confirms what Steno says, that the Strata of the highest mountains are often Simple, without those heterogeneous mixtures of shells, Bones, plants, &c. which abound in the Layers of the lower Grounds.

He added that he had observed similar examples of 'Pikes or fragments' standing on the summits of high mountains in Switzerland.[20]

From Swansea Lhwyd also wrote to John Mill (1645–1707), Principal of Edmond Hall, describing the enigmatic Pillar of Eliseg[21] and enclosing a copy of his transcript of its damaged, and by today illegible, Latin inscription.[22] Robert Vaughan of Hengwrt had transcribed the inscription in 1649, sending two copies (both lost) to Archbishop Ussher and preserving one in his commonplace book.[23] Although Lhwyd had learnt about the pillar from Aubrey in 1693 (G, p. 208), he did not mention it in his additions to *Britannia*. In early 1695 Lhwyd told Hugh Jones of an 'inscription at Lhan Golhen Abbey ... mention'd in one of Ussher's letters', noting that Aubrey had said 'the stone was falln down if not broken several years since' (G, pp. 274-5, date from *EMLO*). John Lloyd sent Lhwyd further details shortly before the tour (G, p. 279), and Lhwyd wasted no time in visiting the site. He told Mill that he hoped Ussher had sent a transcript of the undamaged inscription to Gerard Langbaine (1608/9–58), and asked Mill to help him find it in Langbaine's papers (G, pp. 306–7).

Before the middle of October Lhwyd had returned to Oxford, having 'rambld (very much to my satisfaction) through 8 or 9 Counties' (G, p. 311). He had made many contacts throughout Wales and could better assess how much support he could expect. He told several correspondents about his discoveries. His letter to Richardson 24 November 1696 mainly discussed the new plants he had encountered, though he also noted that his splendid crystal had been found 'above Phynnon Vrech'. This crystal was soon to occupy pride of place as the first entry in the *Ichnographia*. Lhwyd also mentioned a few formed stones, but the close of the letter indicates a shift in his interests: 'I employed the greatest part of my time in copying inscriptions, taking catalogues of Welsh manuscripts, &c.' (G, p. 315). This change in emphasis can also be seen in a letter to John Lloyd written a month earlier: Bishop Humphrey Humphreys of Bangor had been

extraordinarily obliging ... He gave me leave to take a Catalogue of his MSS. which thô considerable enough are yet much inferior to the Collection at Hengwrt ... the most valuable \in its kind/ any where extant: thô I found no Manuscript there which I could safely conclude to have been written five hundred years since. (G, p. 311)

Lhwyd's references to manuscript collections indicate the importance of literary texts as sources for his planned History of Wales. As well as yielding new historical information, these unpublished texts would provide evidence of the oldest form of the Welsh language. Enquiries about manuscript collections increased as work on the Design intensified. Lhwyd knew of the major collections at Hengwrt and Llanforda but had been unable to examine them thoroughly; in Hengwrt he 'took account of as many of ye old Parchment Books' as he could, 'but had not time to run over the paper manuscripts' (G, p. 460). Worse still, his request to borrow four Hengwrt manuscripts he had noted on his first visit was rejected by Howell Vaughan (G, pp. 312–13). His frustration is clear in his comment in *Glossography*, p. 225: 'I have been admitted to a Transient view (for a few hours) of these Collections.'

Lhwyd's quest for Welsh manuscripts was similar to Tanner's larger-scale project, an enlarged edition of Leland's 1546 catalogue of British authors which eventually became the monumental *Bibliotheca Britannico-Hibernica*, posthumously published in 1748. The earliest letters exchanged between Tanner and Lhwyd date from February 1697, but they had known each other since their work on *Britannia*, if not earlier. Before Lhwyd set off on his preliminary tour, Tanner (in a now lost letter) had asked him for information about Welsh writers and manuscripts. On 25 September 1697 Lhwyd drew his attention to the list of Welsh writers in John Davies's *Dictionarium Duplex* (1632) and offered to tell him who now owned the manuscripts (G, p. 347). In his reply, Tanner asked Lhwyd for much more information than Davies had provided. He appreciated that this request was 'unreasonable' but said he would 'always gratefully own from whence I am inform'd, and perhaps one time or another it may lay in my way to serve you in like nature'.[24] Lhwyd sent Tanner a catalogue in April 1698 (G, p. 369), followed by a friendly letter the following month

asking him to note Welsh manuscripts and any relevant Latin texts
he might encounter:

> No news could be more acceptable than your going to London &
> Cambridge, on so excellent a Design. Pray put a litle paper book
> in your left pocket markd N.L. [Ned Lloyd] and as any Lhyvrau
> Kymraeg occurre, think of yr old friend. I desire you would please
> to write ye first & last 4 or 5 words of each Treatise, adding the
> number of leavs \(& lines in a page/) as also whether on paper
> or parchment, and a mark where you think them considerably
> ancient. (G, p. 371)

Lhwyd continued to work on his own catalogue, which was eventually
published as Chapter 7 of the *Glossography*.[25]

A few weeks after Lhwyd's return, Lister told him that Walter
Moyle (1672–1721), MP for Saltash, a botanist and ornithologist, had
suggested that were Lhwyd to visit Cornwall in the summer of 1697 he
would receive '30 subscriptions, besides a hartie welcome'.[26] Although
this tempting offer 'would be most to my interest', Lhwyd rejected it
since he considered himself bound by promises to his subscribers to
make Wales his 'first businesse' (G, p. 314). He had already explained
to John Lloyd how he now intended to pursue his research:

> My Design hereafter is to spend a month or two (according to its
> extent) in each County; and so bid adieu to it. thô I think I have
> taken the best course the first year, to ramble as far as conveniently
> \I/ could in order to inform my self what helps I may expect from
> Manuscripts &c. in general: and to give more general satisfaction
> to the Gentry. (G, p. 311)

By the end of 1696 he was sufficiently encouraged to publish a four-
page pamphlet, *PAROCHIAL QUERIES In Order to A Geographical
Dictionary, A Natural History, &c. of Wales. By the Undertaker E. L.*[27] He
hoped it would receive 'a kind Reception (as from this last Summer's
Travels, I have great Hopes it may)'. The thirty-one queries, sixteen

about 'the Geography and Antiquities of the Country', and a further fifteen about 'the Natural History',[28] based on the 1695 *Design* also included queries on Welsh dialects and the English dialects of Pembrokeshire and Gower. The *Queries* may owe something to Plot's 'Design for a Natural History of Britain', and to other contemporary sets in Lhwyd's possession.[29] Emery considered that the set drawn up in 1677 by the Westmorland antiquary Thomas Machell (1648–78), which was organised topically and sought information at the parish level, was the most influential of these,[30] but whichever models Lhwyd drew upon, his *Queries* appears to have been the most comprehensive of all such compilations. Possibly to encourage respondents, he was the first to provide space beneath each query for their answers, informants being encouraged to continue, if necessary, on a separate sheet of paper. Four thousand copies were printed, three copies to be sent to every parish in Wales for completion by the local incumbents, school masters and gentry. Responses would provide a collection of data on which the study would be based. In fact, nobody answered every query, for they were too wide-ranging to be adequately answered by any individual, and Lhwyd made it clear that he and his assistants intended to visit parishes to gather more information in person. Lhwyd nevertheless sought to ensure that respondents would follow his scholarly standards. They were requested to be concise and confine themselves 'to that Parish only where they inhabit'. They should always distinguish 'betwixt Matter of Fact, Conjecture and Tradition':

> Nor will any, I hope, omit such Informations as shall occur to their Thoughts, upon Presumption, they can be of little use to the Undertaker, or the Publick, or because they have not leisure to write down their Observations so regularly as they desire: Seeing that what we sometimes judge insignificant, may afterwards upon some Application unthought of, appear very useful.[31]

Finally, that Lhwyd was 'Qualified for this Undertaking' was vouched for by four leading scholars with diverse interests, John Wallis, Edward Bernard, Lister and Ray. Lhwyd had originally hoped that Dr Ralph

Bathurst (1619/20–1704), President of Trinity College and one of the founders of the Royal Society, would add his name, but for some reason he failed to do so (G, p. 329). To maximise support and publicity for the project, Lhwyd exploited his extensive network of family, friends and acquaintances as well as scholarly contacts, all of whom were given parcels of *Queries* and of the *Design* to distribute in order to attract subscriptions. His letter to John Lloyd 26 December 1696 gives an indication of the scale of Lhwyd's activity (G, pp. 316–18).

Before Lhwyd could embark upon the *Archæologia* tour he had to address several practical issues. One was finding a trustworthy agent to oversee the finances of the project and to arrange the safe delivery to Oxford of specimens collected on the tour. Lhwyd was fortunate in his choice of agent, perhaps following the advice of Thomas Mansel or John Williams.[32] Walter Thomas of Barnard's Inn, a prominent London solicitor whom he had engaged by early February 1696 (G, p. 298), proved to be an excellent manager who was totally committed to the project; his response, when the agent of a delinquent subscriber 'pretended to Censure the matter & the undertaking',[33] was 'God dammee had he been anywhere Else I would have struck him & I thinke could have brought him to the ground.' He did his best to ensure that Lhwyd's queries were sent to those who would 'not make Bum fodder of them',[34] managed the accounts, dunned recalcitrant subscribers, and, in some sixty letters between May 1696 and April 1704, ensured that Lhwyd was kept fully informed about his affairs.

Assistants had next to be appointed, both at the Museum and to accompany Lhwyd on his travels. Over the years Lhwyd had encouraged Welsh undergraduates to develop an interest in natural history and, when he could, had found some of them a niche in the Museum, much as Plot had taken on Lhwyd himself. He employed some undergraduates as his personal assistants; others were appointed to posts at the Museum, such as Underkeeper or deputy, and were paid from Museum funds when its income permitted. However, as his research became focused on major projects, he realised that he needed a team of assistants to ensure continuity, one of whom would serve as his deputy and, possibly, his successor. His first appointment, as Underkeeper, was probably

William Jones (see Chapter 5), who was succeeded in 1694 by Hugh Jones (1671–1702), a promising naturalist and a fine draughtsman, who left in the autumn of 1695 to become chaplain to the Governor of Maryland.[35] Following his departure, Lhwyd emphasised to Richard Mostyn in November 1695, in a letter enclosing a copy of the proposals, his concern about the need to appoint a potential successor should he die before completing his work:

> There is one very obvious Objection, which I have not taken notice of in the Paper, because indeed I could not well answer it. And this, is that if it please God I should dye before either of these Books be fitted for the Presse; all the encouragement given me, would be so much thrown awa[y]. In order to provide for such an Accident, as well as I can; I shall endeavour to make choice of a young man, of some extraordinary Parts & Industry, for an Amanuensis; and shall instruct him (as far as I am capable) in the Studies of Natural History, and Antiquities; that so he may be qualified not onely to assist me in this Undertaking, if it please God I should live to goe thorow with it; but perhaps to finish it as well or better then my self, if it should happen otherwise. (G, p. 293)

Writing to John Lloyd a week earlier, Lhwyd set out his intention of choosing 'a young man for an Amanuensis, who has parts enough to make at least as good a Naturalist and Antiquary as I am in a few years', adding, 'I shall spare no pains to instruct him' (G, p. 292). In his letter to Mostyn, Lhwyd had said, 'I have already an Eye on one whom I think fit for the Purpose, and also very desireous of such an Employment' (G, p. 293). The young man was not named, but in mid-November Lhwyd told Lister that:

> My Assistant and Fellow-Traveller is one Mr William Rowlands Batchelr of Arts of Oriel College; who is an ingenious fair-condition'd youth and has tolerable skill in surveying and designing; and is \also/ as well acquainted as my self wth the ancient and modern Language of our Countrey; nor doe I despayr

but that in a short time he'll be as well qualified as I am at present to carry on the *Dictionary* and *Archæologia*, if it please God I should dye before I finish them. (G, p. 314)

Rowlands had already provided valuable material for *Britannia*, but although apparently eager to participate in the new project, did not accept the offer, though he subsequently kept in touch with Lhwyd.[36] The search for a deputy and for 'fellow travellers' continued, but several promising candidates were unable to accompany Lhwyd, because of family pressure or the uncertainty of the employment.[37] Early in 1697, William Williams was appointed Underkeeper. Together with Robert Thomas, the Under-Librarian, he was responsible for administering the Museum and keeping in touch with Lhwyd throughout almost all his tour.

By the spring of 1697 Lhwyd had also appointed three assistants to accompany him on his tour. Two were already known to him. William Jones had worked at the Museum from December 1691 onwards; in November 1696 Lhwyd sent him on a fruitless mission to Howell Vaughan at Aberffrydlan to borrow four Hengwrt manuscripts noted by Lhwyd in 1696 (G, p. 313). Robert Wynne (d. 1729) was a member of a prominent Merionethshire gentry family, whose account of the mysterious Merionethshire fires had prompted Lhwyd to describe him to Lister as an 'intelligent sober person' (G, p. 242). Wynne accepted Lhwyd's terms at the beginning of September 1696 and spoke of coming to Oxford in late January.[38] The third assistant, David Parry of Cardigan (1682–1714),[39] was considerably younger than either Jones or Wynne. His father, William, and Lhwyd's 'cousin' John Parry (d. 1727) had recommended him as an assistant (or perhaps pupil) in mid-December 1696, hoping 'you will have him entered in ye universitie as soon as you think itt convenient.'[40] On David's arrival in Oxford, delayed until March 1697 by an attack of smallpox, terms of an agreement were drawn up for 'meat, drink, cloaths, & £5 Salary yearly during his 4 years service', a term subsequently extended to five years.[41]

The group usually worked in pairs, Lhwyd taking Parry under his wing, while Jones and Wynne almost invariably travelled together. All collected specimens, transcribed inscriptions and manuscripts, described and drew antiquities, interviewed informants and compiled glossaries. Lhwyd praised their skills to Thomas Molyneux in June 1698: each had 'some Skill in Plants, but their mean Tallent is in figured Stones wherein they have been very Conversant these two Years'.[42] He had faith enough in his teaching and in their innate ability to allow them increasingly to judge for themselves what was to be recorded and, at times, plan their own journeys. As the most experienced assistant, Jones was *de facto* Lhwyd's deputy and appears to have been responsible for making many of the practical arrangements, such as finding accommodation and dispatching boxes of specimens to the Museum.

The final preparations involved assembling the necessary apparatus. Robinson advised Lhwyd on the practical difficulties of transporting fragile equipment and specimens:

> As for a Microscope the larger Sorts are not tractable in a Journey, are often out of order, and are very dear; therefore I have sent you some pocket Itinerant glasses with the aforementioned Books, such as will discover Animals in fluids, and magnify solid Bodies to that degree, as may be usefull for the designing part, and for enlarging the Seeds, Vessells, &c. of some Sorte of Herbs, Insects &c. ... Spirit of Wine and glasses often miscarry by the Carriers ... but I fancy a good clear filtrated Decoction of Salt in water (Brine or Pickle) will do your Business better than Sp. of Wine, which contracts, shrivells, and sometimes wastes the Bodies. Dr Lister says he has frequently Animals come fresh above 200 miles in these Pickles or filtrated Brines; and I remember they use them much at Leyden.[43]

Lhwyd intended to collect specimens on a very large scale, for in early March Walter Thomas queried his order for '10 Grosses of round boxes, halfe a grosse of square, the round will be 1440 boxes. Pray send word if you intend soe many.'[44] These were presumably the same kind of round boxes for storing small fossils noted by Uffenbach when visiting the Ashmolean in 1710.[45] By April 1697 all was ready, and towards the end of the month the equipment – instruments, microscopes, magnifying glasses, and the 'moving library',[46] Lhwyd's working collection of books – had been packed and sent on to the first staging post.

Most of the notebooks and journals which Lhwyd and his assistants must have kept have been lost.[47] There are references to some nineteen volumes of travel notes, in addition to eight volumes of drawings in the Wynnstay sale. A few chance survivals provide a great deal of miscellaneous information on various locations and provide an invaluable insight into Lhwyd's methods of recording monuments and inscriptions.[48] They indicate generally where Lhwyd and his assistants went, but do not reveal their itinerary (the exception is noted below) and only occasionally the date; for the most part, the route and progress of the tour must be gleaned from the addresses of letters sent and received.

Lhwyd and his companions set out from Oxford in May 1697 and began by retracing the route of the 1693 *Britannia* journey. As Lhwyd had told John Lloyd the previous October, he intended to 'begin in Monmouthshire as being but a day's rideing hence and lying next to Glamorganshire; where the Gentry have subscrib'd as much as a third part of all Wales' (G, p. 311). En route they spent '3 or 4 days in the coal pits of the Forrest of Dean', where Lhwyd found several puzzling fossil plants: 'I know not well whether the impressions of stalks might be easily distinguish'd from those of the Roots' (G, pp. 335–6). They proceeded via Chepstow to Abergavenny, near which place Lhwyd told Robinson that he had discovered 'some new species of *Glossopetræ* and *Siliqastra* (the first *Ichthyodontes*, I suppose, that ever were observ'd in Wales)' in limestone 'on the top of a high mountain called Blorens [Blorenge]', and 'a large Testaceous [hard shelled] body, not to be compar'd as to its figure with any sort of shell yet describ'd' (G, p. 337). After visiting coal and iron mines at Llanelli (Breconshire), Lhwyd described in detail what was clearly a new process to him, coal mining by driving drifts or levels: 'great inroads made into the side of the hill, so that three or four horsemen might ride in abreast. The top is supported by pillars left at certain distances.' He did not collect plant stalks from the 'slat' [shale above the coal] because 'it seem'd impossible to reduce them to their several proper species', but close by the pit we found a valuable curiosity, viz. a stone for substance like those we make lime of; of a compress'd cylinder form ... about 8 inches long, and 3 in breadth; its superficies adorn'd with equidistant dimples, like Dr. Plot's *Lepidotes* ... and in each dimple a small circle; and in the center of each circle a little stud like a pin's head. This is the only curiosity of the kind I have seen; and is not referable to anything I can think of, either in the animal or vegetable kingdom. (G, pp. 337–8)

Plot had published in his *Oxford-Shire* a description and illustration (Plate 3, item 11) of a non-Oxfordshire fossil which he considered to be the skin of a carp or similar fish. Gunther identified Lhwyd's find as *Lepidodendron* (G, p. 337), a massive Carboniferous-period arborescent lycopod, one of the dominant plants of the coal swamps, but North's earlier identification of it as *Stigmaria* would appear to accord better with Lhwyd's description and Plot's illustration.[49] In a sense both were correct since *Stigmaria* is the basal rooting organ of *Lepidodendron* and *Sigillaria*, and was considered to be a separate plant until the mid-nineteenth century.[50]

By 6 June Lhwyd was in Pontypool, one of the few precise dates for his journey, since he published an account of the exceptionally violent hailstorm he experienced there. Some of the hailstones were 'eight Inches about', crops were beaten down and four pounds worth of glass in Sir John Hanbury's house was damaged.[51] He was impressed by Hanbury's 'excellent invention', a machine for rolling hot iron into thin plates for manufacturing 'furnaces, pots, kettles, sauce-pans, &c ... at a very cheap rate' (G, p. 338). Lhwyd's interest in this venture is a reminder that seventeenth-century naturalists were interested in the practical applications and economic value of their discoveries. Thus in April 1699 he told Richardson that he had discovered 'a sort of Marble ... which when polishd represents a number of small Oranges cut a cross ....This might serve very well for inlaying work, as Tables, windows, cabinets, closets &c. & would make curious salt sellers.' He added 'If you are acquainted with any gentlemen that deal in Alome or Coppras, you may please to acquaint [the]m that Wales affords good Quantity of each, if they judge it worth their while to put up any works there' (G. p. 415). The party pressed on through Usk to the Vale of Glamorgan, reaching St George (18 July) and St Nicholas (22 July). In Cowbridge, 22 September, he told William Williams that he planned 'to winter near Cardigan; and to survey that County and Pembrokeshire if Possible before spring'. He had intended reaching Carmarthenshire 'about a month since', but had spent two months copying 'an old Welsh MS.', borrowed by Thomas Wilkins (1625/6–99) of Llanfair, which Lhwyd might have bought for twenty shillings had Wilkins returned it to its owner (G, pp. 343–4). He described it to Robinson as 'a large Welsh MS. ... writ on velome, about 300 years since; and contained a collection of most of the oldest writers mentioned by Dr. Davies at the end of the *Welsh Dictionary*'.[52] This was the Red Book of Hergest, now Jesus College MS 111. Lhwyd also noted recent discoveries of plants, butterflies, fossils and antiquities, including 'a Flint Axe, somewhat like those used by the Americans', an example of Lhwyd's readiness to combine archaeology with anthropology. He had visited Goldcliff, where he found formed stones and an *Asteria* 'beset with Sprigs ye whole length of it', Barry Island (large *Entrochi*) and Caerffili Castle, where he noted a possible inscription and masons' marks (G, pp. 344–6). By October the group had reached Swansea and was ready to proceed to Carmarthenshire.

The 'Miscellaneous Memoranda of Wales out of E. Lhwyd's MSS, the Panton Collection &c &c.', Cardiff Free Library MS. 2.59,[53] appears to be field notes compiled by Lhwyd and his assistants on their visits to twenty-four Carmarthenshire parishes. These were later copied for the Pembrokeshire antiquary Richard Fenton (1747–1821), the copy unfortunately omitting drawings and reproductions of inscriptions.[54] The order of the parish descriptions in the 'Memoranda' may reflect the route followed by the group (or in pairs), westward along the coast to Pen-bre and north to Carmarthen, where on 24 November Lhwyd wrote to the French botanist Joseph Pitton de Tournefort (1656–1708) (G, p. 351), then on to Meidrim, Tre-lech, Cynwyl Elfed and Cenarth, before completing the circle in the north-eastern parishes from Llanllwni to Cil-y-cwm and Llanddarog. The 'Memoranda' does not include all the parishes visited on the tour, for Lhwyd's correspondence shows that Cydweli and St Clears were also included. From Llandeilo, towards the end of December, Lhwyd sent John Lloyd an account of 'this year's ramble' through Monmouth, Glamorgan, Carmarthenshire and Cardiganshire. Discoveries included 'several remarkabl Form'd Stones on the Shoars and in the Quarries', but Lhwyd had not 'seen one *Belemnites*; whch you know is the most common about Oxford, and indeed \in/ all those parts of ye Island from the Severn shoar to ye remotest parts of Sussex and Kent'. He had 'met with several Welsh MSS. but not above 2 or 3 of any considerable Antiquity', though he had transcribed 'a fair large folio \on velom/ containing copies of such old MSS. as ye Writer could meet with', the Red Book of Hergest noted above. He also noted inscriptions and crosses in Margam, Ewenni, Carmarthen and Dinefwr. Finally, he told Lloyd in confidence that his 'expences can not be lesse than £15 per annum'.[55] At the end of February 1698, Lhwyd described to John Lloyd in some detail his reading of the poems in the 'very ancient & Valuable' Red Book associated with Llywarch Hen (G, pp. 355–6). Lhwyd was later to persuade himself that one of these, 'Marwnad Gereint ap Erbyn', was a specimen of ancient Cornish. Amongst the fossils, he noted and illustrated the trilobite discussed in Chapter 6.

The group was now ready to winter in Tenby, Scotsborough House providing a convenient base for excursions into Carmarthenshire and south Cardiganshire. In January 1698 Lhwyd visited his mother's cousin, John Parry of Panteinion and Pant-yr-odyn, rector of Trefdroyr (Troed-yr-aur) since 1680. An enthusiastic antiquary, Parry had subscribed to the *Glossography*, distributed copies of the *Queries* and collected subscriptions.[56] Lhwyd's Cardiganshire relatives knew of his travel plans and looked forward to welcoming him. When he heard of this visit to Parry, Lhwyd's uncle, Edward Pryse of Gogerddan, was vexed that Lhwyd had not visited his Pryse relations.[57] In Tenby, Lhwyd wrote *Epistolæ* I, II and III of the *Ichnographia* in March, April and May 1698, and in April sent Tanner the 'Catalogue of Welsh authors' he had agreed to prepare for him the previous September (G, pp. 369). The group also began exploring Pembrokeshire, possibly seeking 'sea plants' for the Duchess of Beaufort (G, p. 358). On 22 March Robert Wynne sent Lhwyd an illustrated account from Haverfordwest of discoveries he and William Jones had made at Llangwm, where 'Mr. Bowen was exceedingly civil to us'. After examining the shore at Milford, where their attempts to dredge for specimens was defeated by high winds, they landed at Burton, then 'marched on from parish to parish (sometimes meeting with tolerable estempore information) … til we came to St Brids'. They proceeded via Marlais to Haverfordwest, where they met Lhwyd's correspondent, John Pember (1664–1735), and intended to proceed to St Davids. By 8 April they were in Letterston, having explored Dewisland and St Davids, where they had been impressed by 'Monuments which are too numerous to be all mention'd & so equally valluable yt none claym's a priority'.[58] These letters show Lhwyd's assistants maturing into trusted investigators and also the extent to which the project relied on the cooperation of local clergy and gentry. It appears that Lhwyd was at Pembroke 20 May, where he wrote an account of hunting the wren and other folk customs,[59] followed by copies of inscriptions at St Davids cathedral. There followed visits to Haverfordwest 21 May, Llawhaden 10 June, with descriptions in Latin of seven butterflies, headed 'Papiliones Glam. Pelleiod',[60] 'prope Henri's moat' 13 June (notes on Pembrokeshire words), and Narberth 19 June (G, p. 370). From Pembroke in early June Lhwyd sent Lister an illustrated account of his fossil finds:

> These are scarce a tenth part of our discoveries in this kind, the last year & the present; but some of them being *toto genere* new; and \the rest/ improvements of your own & other Gentlmen's observations, I presum'd this faint representation of them would be some Diversion to you. (G, pp. 399–400, date from *EMLO*)

Moving north to Ceredigion, Lhwyd had reached Cardigan by 6 July (noting some interesting folk customs) and Lampeter, whence *Epistola* IV was dated 15 July (G, pp. 379–80). He then travelled to north Wales through the Marches and mid-Wales. Lhwyd wrote *Epistola* V to Ray from Rhaeadr 29 July (G, pp. 381–8). Writing to Richardson from Hay 19 September, Lhwyd discussed the plants that he had observed; though most were familiar, a few were new. He added that 'In a \great lake/ called *Lhyn Savadhan* I found a pellucid plant I had never met with before ... We found there also ... two elegant sorts of small Leeches, which I suppose not describ'd' (G, p. 401). He was particularly impressed by another feature, the 'Fayrie Causways', discussed in the following chapter.

By 18 October Lhwyd had reached Newtown, probably via Llanbadarn Fynydd since a letter with that date was written there. The following day he was in Montgomery, 'at Mr Thomas Francis a painter' (G, p. 402). This was Thomas Francis (?1621–1700), a professional heraldic artist and a deputy herald for north Wales since 1664.[61] Here Lhwyd received a letter from Richardson which enclosed two plates of fossil plants and 'insects' for the *Ichnographia* (G, pp. 402–4, with reproductions). It was here, too, that he composed the Preface to the *Ichnographia*, dated 1 November, but by late November he had moved on to Gwersyllt near Wrexham (G. p. 404). Meanwhile, by late October William Jones had reached Dolgellau, where he was 'civilly entertained' by Lhwyd's old friend, Maurice Jones, who accommodated him at the rectory, and would 'not suffer me to go to my quarters till the beginning of the n[ext] month'. Maurice Jones had arranged accommodation for the group

> at Mr Robts the ministers mother ... a house ...of very good & honest reputation and extraordinary civil to all sorts of people; and the chamber tho small hath fire place in't, and out of the noise of [the] town very convenient I think to our purpose being furnished with all necessarys as \a/ Bed &c. Mr Roberts doth truly offer the benefit of the chamber at free cost.[62]

The group wintered in Dolgellau from December 1698 to March 1699 before continuing their journey northwards.

Some of Lhwyd's field notes are preserved in two surviving notebooks. NLW, Peniarth MS 251 is an interleaved copy of Ray's 1696 *Synopsis Methodica Stirpium Britannicarum*, presented to Lhwyd by the author. It contains miscellaneous notes, many dated 1698, on the antiquities and natural history of south Wales (with some drawings and inscriptions) and descriptions of some north and mid-Wales parishes, mostly in Lhwyd's hand.[63] NLW, Llanstephan MS 185,[64] dated 1698, contains memoranda – drawings, inscriptions, sepulchral slabs and antiquarian notes, examples of local dialects, some parish descriptions – compiled at various times from a number of areas. These notebooks provide much additional information about places visited, together with descriptions of discoveries made there. Although they provide a valuable indication of the general direction of the tour, they do not enable one to reconstruct a detailed itinerary. According to Peniarth 251, the tour included the Breconshire parishes of Llanhamlach (1 September); Llansbyddyd, Llanfyrnach, Llan-y-wern, Crucadarn, Crucywel, Talgarth, Llangatwg and Hay (19 September); the Herefordshire parishes of St Michael's Escley, 'Kapel Koed Klavyl … ym mhlwy trewalter [Walterstone]' and Cusop; the Glamorgan parish of Gelli-gaer and the Monmouthshire parish of Bedwellte (29 August). There are parish descriptions of Llys-wen and Aberllynfi (Breconshire); Glasbury (Brecon and Radnorshire); Colfa, Glasgwm, Llanfihangel Nant Melan, Bleddfa(ch), Casgob, Pilleth ('Pillis yn Saesneg, Pillalley yn gymraeg'), Llangynllo, Bugeildy, Heyope, 'Lhandhewi yn Hiop' (Radnorshire).

The notes in Llanstephan 185 are more haphazard; they are listed here by county while retaining the order in which they are noted in the book. In Carmarthenshire, Llanboidy; in Breconshire, Defynnog, Merthyr Cynog, Llyn Syfaddan, Llanelli, Llanddeti, Llangatwg, Llanwrtyd, Llanllywenfel (Llanlleonfel), Llanafan Fawr; in Montgomeryshire, Kerry, ('Lhanhiangle Church'), Welshpool ('Memorand[um] yt upon ye 21th day of 7ber 1698 I weighed ye Gold Communion Bowl of Pool Church'); in Shropshire, Clun; in Glamorgan, Barry Island, Merthyr Tudful, Cowbridge; in Radnorshire, Llanfaredd, Llowes 'Llywes Church yd.', Diserth (with a fragment of Welsh folklore); in Herefordshire, Llanfeuno, Ewias Lacey; in Monmouthshire, Llanwenarth. The Defynnog entry is of particular interest since it includes an illustration of the now lost Llywel stone bearing roman-letter and ogam inscriptions. Although Lhwyd recognised ogam as a form of writing and discussed it in the *Glossography*, the script puzzled him.[65]

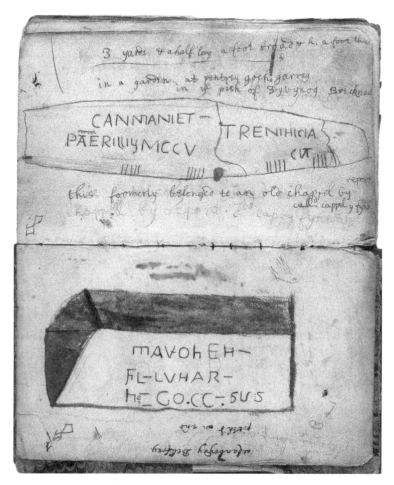

FIGURE 7 Two pages from Lhwyd's only surviving field notebook. The upper page depicts the lost Crai (Llywel) stone with Latin and ogam inscriptions, the lower, a now fragmentary stone from Llanboidy recording the killing of Mavohenus

The contents of the parish descriptions in the 'Memoranda' and in these two notebooks follow the sequence of the headings in the *Queries*, which was apparently used as an aide-memoire on the tour. Both notebooks remained in use as the tour progressed to north Wales from Dolgellau through Denbighshire and Merioneth. Peniarth 251 contains a number of parish descriptions, which may provide a general indication of the route taken through Llangollen, Corwen, Llansanffraid Glyndyfrdwy, Gwyddelwern, Llangwm, Llanfor, Llangar, Llandrillo, Llandderfel, Llanfor (a second account), Llanycil, Llangywer, Llanuwchllyn and Betws Gwerful Goch. Llanstephan 185, in addition to the references given above, records material, inscriptions, plans and drawings: from Montgomeryshire, Kerry, 'Lhanhiangle C' (possibly Llanfihangel-yng-Ngwynfa), Llanidloes, Trecastell (Llanwnnog); and from Merioneth (Trawsfynydd). The manuscript is here turned over and entries recommence from the end. The first is the note about Lhwyd's birth quoted in Chapter 2, followed by some dialect words, mainly from Montgomeryshire, and descriptions of the Merionethshire parishes of Llanfihangel-y-Traethau, Llandecwyn, Maentwrog and Llanfrothen.

During the winter months at Dolgellau, Lhwyd corrected and revised the *Ichnographia* as it went through the press. He did not see the final proofs and, writing to Lister on 18 January 1699, expressed his anxieties about the book's reception since it was 'so far out of the common road' (G, p. 412). A month later he thanked Lister for his 'noble present' of the original copper plates, asking him to send them to the Museum (G, p. 413). Were he provided with impressions of those plates which he lacked, Lhwyd could 'patch up' a copy for his own use in correcting the errata for a prospective second edition, thus freeing up a copy for some local subscriber 'who perhaps might procure me two or three others'. He was still in Dolgellau when he wrote to Humphrey Foulkes mid-March 1699, noting 'Tlws Owen Kyveiliog'. At Glascoed he had seen 'an excellent collection of Welsh papers but not near so many parchment MSS as at Hengwrt',[66] and commented on a puzzling inscription at Tywyn, 'of five lines, which by ye form of the letters I guess might have been of ye 7th or 8th century, but, tho the letters be very plain – I can make neither Welsh nor Latin of it', a reference to the important 'Cadfan' stone, the only Dark-Age inscribed stone bearing a Welsh inscription.[67] On 18 April he sent Richardson a description and illustration of 'a remarkable Sea Plant' found while dredging for oysters near Llandanwg (G, pp. 414–15). This letter was written from Gogerddan, where Lhwyd had overwintered two horses.

His uncle, Edward Pryse, told Lhwyd 1 April 1698 that, though well cared for, they had contracted farcy, and Lhwyd was to collect them to prevent them infecting other horses. Lhwyd took advantage of the journey to revisit Llanddewibrefi, since his failure to read the whole of the IDNERT inscription in 1693 had continued to vex him. On this visit he removed the stone from the wall, reporting the complete inscription and his corrected interpretation in a letter to Bishop Humphrey Humphreys from Dolgellau 29 March 1699; a fuller account appeared posthumously in the 1722 *Britannia*.[68]

Lhwyd continued his tour, writing from Tal-y-cafn in the Conwy valley 25 May and from Flint 26 June, but Ireland and Scotland were now increasingly occupying his mind as he made his way, by 1 August, to Beaumaris in Anglesey to cross to Ireland. Before embarking on the next, more hazardous part of his tour, Lhwyd appears to have sent some, at least, of his books back to Oxford, Tanner telling Lister on 13 November that 'about six weeks since the Welch Carrier brought up his moving library and several Boxes of Stones & other Collections, and said that he left him about the latter end of August going on Shipboard for Ireland.'[69]

ARCHÆOLOGIA BRITANNICA:
'PARALLEL OBSERVATIONS'

Ireland: August–November 1699

Lhwyd's 1695 *Design* had stated that the Welsh tour would be supplemented by visits to Cornwall and Ireland or the Highlands of Scotland
for making 'parallel Observations'.[1] When he sent a dozen copies of
his 'Quæries' to Thomas Molyneux in Dublin in June 1698 he said
that he and his assistants intended to 'ramble through such places of
the Kingdom as you shall direct for the Space of at least 2 Months'.[2]
Lhwyd's Irish tour is, as Campbell suggests, best considered as a single
journey interrupted by a three-month visit to Scotland.[3] Lhwyd had
visited Ireland briefly in 1688, had begun to learn Irish, had a little
experience of the spoken language and, crucially, had personal contacts
there. He and his companions sailed to Ireland towards the end of
August 1699, having been reassured by Thomas Molyneux in May 'yt
you may Travell ye Roads of Ireland as safe as ye Roads of England'.[4]
Once in Dublin, Lhwyd spent three days meeting friends and support-
ers and listing Irish manuscripts in Trinity College Library,[5] before
travelling north to inspect the Giant's Causeway, a 'place I reckon
well worth my Journey, whatever we meet with besides'.[6] This area of
some forty thousand interlocking pentagonal and hexagonal basaltic
columns up to ninety feet high, formed by the cooling and cracking
of ponded lava some sixty million years ago, had been drawn to the
attention of natural philosophers by Sir Robert Redding (d. 1689) in
a letter read to the Royal Society in January 1689.[7] Although this

remained unpublished, it inspired Richard Bulkeley, FRS (1660–1710), to publish a description in the *Philosophical Transactions* for April 1693.[8] Samuel Foley (1655–95), the bishop of Down and Connor, and Thomas Molyneux responded in the August 1694 issue of the *Transactions*, Foley providing a fuller account of the columns, and Molyneux correcting errors in Bulkeley's description and offering suggestions as to the origins of so 'Remarkable a Natural Curiosity'.[9] Molyneux particularly wished to 'obviate a gross mistake' by those 'perfect strangers to Natural History' who believed that the Causeway was 'the Workman-ship, of Art and Mens Hands, [rather] than an Original Production of Nature'. He argued that one could not 'imagine that *Men* could have the least Design in putting all this useless Lumber ... in so Remote and Desolate a Place'. The 'Superstitious People of the Country ... who through Ignorance' had ascribed it 'to the working of *Giants, Fairies, Dæmons,* and such like Imaginary Causes did not deserve any Answer'. The columns, 'the Architecture of the Regular Hand of Nature', resembled gigantic 'star stones' (crinoids). The best analogy he found was the 'Lapis Basaltes Misenus', pillars in Meissen described by Johannnes Kentmann (1518–74) in an account included in Gesner's *De omni rerum fossilium genere* (Zürich, 1565). These 'Misnean Basaltes' appeared so similar to the 'Irish Basaltes' of the Giant's Causeway that Molyneux proposed calling the latter '*Lapis Basaltes, vel Basanos Maximus Hibernicus*'. Although still unable to visit the site himself, Molyneux expanded and improved his description in a second paper in the *Philosophical Transactions* for June 1698, based on new observations and including drawings by Edwin Sandys (d. 1708) that appeared to confirm his earlier account.[10] Molyneux's description would remain the authoritative account until studies by Nicholas Desmarest (1725–1815) in 1771 and Sir William Hamilton (1730–1803) in 1778 confirmed the volcanic origin of these basalt columns.[11]

Even before the first account of the Giant's Causeway had been published, Lhwyd's interest in what appeared to be enormous fossils had been kindled. He wrote to Ray in July 1690 of a 'current report' that in Antrim 'there are divers large pillars of star-stones able to support a church' (G, p. 103). In May 1693 he told Lister that he was 'apt to belive there's some curiosity in ye Irish pillars' but that 'If they are onely elegant pillers and somewhat regular; worne either by the Sea Water; or frequent Storms and Grayheaded Time; I have seen such in Wales' (G, p. 186). By July 1694, after reading later accounts, he recognised the Causeway as 'one of ye noblest Curiosities this age hath discoverd in ye Mineral Kingdom' (G, p. 239). Writing to Molyneux in 1698 he explained its significance for his theory of fossil origins:

> if your Basaltes be as referrable to any sort of Animal as the Entrochi and Asteriæ are to the Vertebræ of Sea Starrs, it putts the Question out of Dispute that Marine Fossills are not the Effects of any Deluge but Mock-Shells, Mock-Teeth &c. formd in the Earth ... If the Bassaltes differs in substance from other Stones its a Suspition it may also be of Animal Descent.[12]

Having viewed the Causeway Lhwyd realised that there was nothing unusual about the composition of its stone; as he told Robinson in December 1699, 'we have the same Stone on the top of Cader Idris ... but ours is less elegant and does not at all break off in joynts' (G, p. 422). Even before Lhwyd had visited the Causeway it had prompted him to coin a new term; in September 1698 he told Richardson that 'The most considerable Rarities [the local limestone] afford are Fayrie Causeys, which I call soe in imitation of their Giants Causey in Ireland ... theirs may be half a mile long; ours seldom exceed three Foot' (G, p. 401). Lhwyd made an analogy between the polygonal calices of basaltiform corals such as *Lithostrotion* and the columnar jointing of the Giant's Causeway.[13]

Lhwyd and Parry (Wynne possibly joining them later) made their way to Antrim via Swords, Dundalk, Armagh, Lisburn, Lurgan, Ballynure and Larne. Lhwyd noted en route a few Irish phrases useful for travellers, some of which he later glossed with their Scottish Gaelic equivalents.[14] He recorded a few monuments and antiquities, notably 'a Copper Trumpet like a Sow-Gelders Horn ... above 2 foot long',[15] three of which had been found 'in an old Karn ... at Balle Niwr [Ballynure] near Carreg Fergus' (G, p. 422). In Armagh he visited his old acquaintance Archbishop Narcissus Marsh and listed his Irish manuscripts.[16] The most poignant encounter was near Larne where

> we met with one Eoin Agniw [Ó Gnímh], whose Ancestors had been Hereditary Poets, for many Generations, to the Family of the O Neals; but the Lands they held thereby being taken away from his Father, he had forsaken the Muses and betaken himself to the Plow: so we made an easy purchase of about a douzen ancient Manuscripts on parchment. (G, p. 423)

The fate of the Ó Gnímhs was typical. The exile of the native Irish landed aristocracy, from the Flight of the Earls in September 1607 to the final 'Flight of the Wild Geese' after the fall of Limerick in 1691, fatally weakened the traditional Gaelic bardic order by depriving it of its patrons. By the time of Lhwyd's tour many, if not most, owners of Irish manuscripts could neither read nor appreciate what they had inherited. By buying such manuscripts cheaply Lhwyd ensured their survival.[17]

On his journey northwards Lhwyd was impressed by a 'most remarkable Curiosity … a stately Mount at a place called New Grange near Drogheda; having a number of huge Stones pitch'd on end round about it, and a single one on the Top' (G, p. 421). This Neolithic passage tomb,[18] the principal feature of the complex of monuments at the bend of the river Boyne, now a World Heritage Site, had long featured in Irish mythology. At the time of Lhwyd's visit, workers employed by a local landowner to remove stones from the mound had uncovered the 'Door of a Cave' (G, p. 421). Lhwyd wrote at least three accounts of the monument,[19] the fullest being that sent to Henry Rowlands (1655–1723):

> Before the entrance we found a great flat stone … placed edgwise, having on the outside certain barbarous carvings, like snakes encircled, but without heads. This entry was guarded all along on each side with such rude stones pitch'd on end, some of them having the same carving, and other vast ones laid a-cross these at top. The out pillars were so close press'd by the weight of the Mount, that

they admitted but just creeping in, but by degrees the passage grew wider and higher till we came to the cave, which was about five or six yards height. The cave consists of three cells or apartments … in the right hand cell stands a great bason of irregular oval figure of one entire stone … within this bason was some very clear water which drop'd from the cave above … Many of the pillars about the right hand basons were carvd … but under feet there were nothing but loose stones … and amongst them many bones of beasts and some pieces of deers horns. (G, pp. 429–30)

The excellent plan,[20] sometimes attributed to Lhwyd himself, was drawn by William Jones, his most competent draughtsman, Lhwyd telling Thomas Molyneux in May 1700 that 'Will Jones gives you his humble Respects and sends you a Draught of the Cave at New Grange.'[21] Although varying in detail, Lhwyd's letters and the plan provide a clear account of the first exploration of the monument. Since a Roman coin 'of Valentinian' was found near the top of the mound, Lhwyd concluded on stratigraphic grounds that the edifice must be earlier than any Danish invasion; his argument, embodying what is today a basic archaeological principle, was totally lost on Thomas Molyneux.[22] The 'barbarous' nature of the 'carving and rude sculpture' proved the mound was 'some place of sacrifice or burial of the ancient Irish' (G, p. 422). Writing to Molyneux in January 1700, Lhwyd added a 'Memorandum' noting that Charles Cambell, the second man to enter the 'cave', had told him that he found human bones there. Since the 'cave' had no 'vent to Discharge the Smoak of a burnt Offering', Lhwyd concluded that 'it must certainly have been the sepulcre … of some person of Note … and not a place for sacrifice.' The animal bones

seem to I[m]port that the person here buryed had been … a great Lover of the Chase and a mighty hunter of these sort of Deer and as a Memorial … their horns were interrd along with him, as we see it is the fancy of some in our own Days to desire that such things they delighted much in when alive should be Buryed with them.[23]

Scotland: November/December 1699–Late January 1700

Although Lhwyd was familiar with the work of Scottish scholars such as Sir Robert Sibbald,[24] he had not met or corresponded with any of them. He had never visited the country, nor had he any experience of Scottish Gaelic. In 1697, in a now lost letter, he contacted the Revd John MacQueen (1643–1733), hoping he could respond to some of his queries and find a Gaelic-speaker to assist him in his researches. MacQueen's reply was disappointing:

> It is long since that I came from the Highlands and I was then very young: I have forgot much of their language and of their Customs ... I never learned their language by book, nor studied to accomplish my self in it, for it served for no use to me, since I intended not to Return to the place where it was commonly spoken, and the truth is that I have a greater itch of Curiosity after it now than formerly I had. Nor do I think it so contemptible as many others do.[25]

Nevertheless, he offered to write on Lhwyd's behalf to Sibbald, 'a worthy ingenious gentleman, my intimat acquaintance' and to his 'many blood Relations and acquaintances' in the Western Isles and Highlands, 'who may be serviceable to you'.

Several of Lhwyd's friends had attempted to dissuade him from venturing to the Highlands, Robinson warning him in April 1699: 'I am afraid you will not be able to make any progress among those Mountaineers, who never or very seldom see any Strangers within their Land.'[26] In fact, Lhwyd was pleasantly surprised by his reception, telling Richardson in December 1699 that 'In the High-lands we found the people every where civil enough; and had doubtless sped better as to our Enquiries, had we the language more perfect' (G, p. 423). In January 1700 he told Thomas Molyneux that in the Highlands he found 'the Gentlemen men of good Sense and Breeding, and the Commons a subtill Inquisitive People & more civill to Strangers in Directing them the way ... than in most other Countys'. Lhwyd ascribed the 'Barbarous' reputation of Highlanders to the 'Roughness of their Countrey ... and their Retaining their Antient Habitts [clothing] Custom & Language. On which very Account many Gentlemen of good sense in England esteem the Welsh at this Day Barbarous'.[27]

Lhwyd and Parry finally embarked for Scotland, probably from Ballycastle, to Southend in Kintyre, in November or December 1699; they intended to stay about two months. Lhwyd's notebooks and other papers which were later donated to Trinity College Dublin enable us to trace his route in Scotland in greater detail than for any other part of the *Archæologia Britannica* tour.[28] He and Parry travelled north along the west coast of Kintyre through Knapdale and Lorn(e), crossing from Kerrera to Mull and then to Iona. The weather was so rough that even Lhwyd had to curtail his botanising: 'Going up one of the High Hills of Mul ... besides some Frost and Snow the wind was so strong that we returnd before we went half way.'[29] He saw a few plants that were new to him, especially on Iona and Mull, where he noted the Bearberry (*Arctostaphylos uva-ursi*).[30] They returned through Argyll to Glasgow via Inveraray, where they met the Revd Colin Campbell of Achnaba (1644–1726), who promised his assistance. In Glasgow, 'the Principal of the College shew'd us several stones he had lately procured ... having Roman Inscriptions', which Lhwyd copied (G, p. 425). By 13 November they were in Edinburgh, where Lhwyd was 'kindly entertaind' by Sibbald and the botanist and numismatist James Sutherland (*c*.1639–1719), who showed him their impressive collections (G, pp. 418–19). Lhwyd transcribed a fair number of inscriptions but 'could meet with no Antiquary hitherto ... that could interpret those in the Irish' (G, p. 429). He told John Morton that he had discovered a monument

within four miles of Edenburrough, which I take to have been of a K[ing] of the Picts ... a circle of rude stones layd flat in the ground, having one very large \one/ pitchd on end, and on that, an Inscription in the Characters usd towards the fifth and sixth centuries: *In oc tumulo iacit Vetta F. Victi.* This \had/ escapd unobservd by Sir Robert & their other Antiquaries; but by this time I suppose they have dug through it.[31]

This was the so-called Catstane ('Cat-Stene') near Kirkliston, Lhwyd being the first to describe the monument and transcribe its inscription.[32] He correctly attributed the 'barbarous' lettering to the 'fifth and sixth centuries' but was mistaken in his belief that the language was Pictish, since it may have been a monument of the British Gododdin tribe, which reused a Bronze Age burial site. Lhwyd's discussion of the name 'Vetta' on the inscription reveals how his linguistic thinking was developing and influencing his historical ideas: 'I suspect the person's name was Getus, of which name I find three Pictish kings; for the names pronounced by the Britains with G were written in Latin with V' (G, p. 429). More recently it has been suggested that 'Vetta' was a woman's name.[33]

By 15 December Lhwyd was in Bathgate, near Linlithgow, where he finally caught up with his correspondence, apologising that he had been remiss in sending reports of his 'rambles of late through countreys so retir'd, that they affoarded neither post nor carrier' (G, p. 418). By 20 December he had returned to Glasgow, intending to return to Ireland via Greenock and thence to Campbeltown, but unfavourable winds detained him for 'about 5 weeks at the Mull of Cantire' (G, p. 426). Although he told Lister that this was 'a place where we could find Litle to doe', his letter to Henry Rowlands, 12 March 1700, shows that he made good use of the enforced delay: 'I have fill'd about three sheets of paper with their [Highlanders'] customs ... and have translated Mr Rays Dictionariolum Trilingue[34] into their language [a south Argyllshire dialect]' (G, p. 428). The 'three sheets of paper' appear to have been preserved in one of his note books, Bodleian MS, Carte 269, a copy Lhwyd made, with annotations in Welsh, of James Kirkwood's *A Collection of Highland Rites and Customes*.[35] Lhwyd also commented on place-names, correctly observing: 'I know not whether the lowlands of Scotland may not agree more with the British than the highlands.'

Lhwyd found a number of charms or amulets in Scotland, including 'a Flint Arrow head set in silver and worn about the Neck. These are the same as the Natives of New England head their arrows with.'[36] Other amulets consisted of fossils.[37] He was particularly interested in so-called 'adder beads' or 'snake stones', which he had characterised in *Britannia* as relics of druidic belief.[38] As a sceptical naturalist he could not accept the popular explanation of how they were formed. Popularly believed to be found only in cocks' knees, the stones were typical of chalk ground:

> so I could never find them in all my travels but at that place, from whence in the time of Paganism the Druids procur'd them and sold them amongst our Northern Britons for Stones of Miraculous efficacy against perils by fire and water; perswading the vulgar they were generated in *Cocks' knees*, as thousands in the Highlands believe at this day. And one fellow had the impudence to tell me (finding me a little hard of belief) that he himself had taken one ... out of a Cock's knee with his own hand. (G, p. 464)

Even so, since Lhwyd thought that the belief might be of value as a lingering relic of druidic rites, he gathered

> a tolerable collection, picked up in Wales by the name of Glain Neidr, in the Highlands by the name of *Crap an Aithreach*, in the Low Lands by that of *Adder-Sten*, and in Cornwal *Milpreve*. These are as celebrated amongst the vulgar in Scotland as in Wales; but in England there is no talk of them, excepting in the west of Cornwal. I am fully satisfied from Pliny's account of the Ovum Anguinum, that these were also Druid Amulets, and am apt to suspect that they had even in those barbarous times, the art of making and staining glass; and that that was the art called Celfyddyd Fferyllt [Art of the Chemist], which Dr. Davies in his dictionary[39] interprets Chymia. (G, pp. 464–5)

These stones and the associated beliefs and customs, he thought, were evidence for a common religious or cultural history represented by druidism. The location of finds demonstrated how widespread that religion was, apart from in Ireland, where, since 'there were no snakes, they [the druids] could not propagate it' (G, p. 430).

Oral and traditional culture and beliefs were an important part of the *Archæologia* project and, as Campbell and Thomson make clear, several Gaelic scholars in Argyll assisted Lhwyd.[40] In December 1699 Lhwyd sent Colin Campbell a list of headings, similar to that used in Wales, indicating the information that he sought.[41] Lhwyd explained what he meant there by a 'translation' of Ray's *Dictionariolum*: he wanted 'An Interpretation of ye Nouns ... with ye Addition of ye Verbs and Adjectives in ye vulgar Nomenclatura, into your Western Ersh'.[42] As shown above, Lhwyd undertook this task himself; the source of some of his Argyllshire words may possibly have been 'the raw Scottish lad', a native speaker whom Lhwyd brought from north Knapdale in the hope of making him an amanuensis and general servant.[43] He was 'a deserted infant [i.e. a minor], whom he brought with him to Oxford, and kept as a servant; his name was Mac Mullein, but they called him Gilia Cholum.'[44] Though the attractions of Oxford were to prove too great a temptation for 'Gilli', as Lhwyd's companions affectionately called him, Lhwyd persevered despite his having to impose many sconces (small fines) on him for offences such as being absent all day, not returning from errands, failing to light candles and staying out all night.[45] An indication of Lhwyd's ambitious plans for Gilli are draft letters and transcripts in the latter's hand bearing his signature, 'Giliecholum McMulen his book Ogust 702', in BL, Add. MS 15072, a notebook containing a catalogue of crystals and fossils, which also notes fines for failing to 'write one page in the space of five weeks' and to transcribe part of a Latin–English vocabulary.

Lhwyd's primary and most valuable source of information on Scottish Gaelic language and culture was John Beaton (*c*.1640–*c*.1708/15), Episcopalian minister of Kilninian in Mull, whom he probably met in Coleraine in February 1700. Beaton, the last learned member of his family, formerly hereditary physicians of the Lords of the Isles, was 'one of the last living persons who had had the ancient Gaelic bardic education',[46] a body of traditional knowledge in medicine, manuscript collections, tales, poetry, traditional or legendary history and national lore. Beaton gave Lhwyd descriptions of those Gaelic manuscripts in his possession, including a collection of poetry, bardic grammar, histories and tales, *Seanchas* ('ancient lore'), laws and medicine, material that was characteristic of the culture of the learned orders.

Lhwyd realised the value of Beaton as a representative of classical Irish/ Gaelic culture and attempted to preserve an example of the archaic nature of learned Gaelic by 'recording' in his own phonetic script Beaton reading the first two chapters of Genesis from Kirk's 1690 Irish Bible.[47] As with the Ó Gnímh manuscripts, Lhwyd's meeting with Beaton was providentially timely.

Ireland: January–August 1700

William Jones and Robert Wynne had been busy while Lhwyd and Parry were in Scotland. In an affectionate letter in Welsh dated Dublin 6 December 1699, Jones sent Lhwyd a lengthy account of his peregrinations through Connaught.[48] He had travelled to Athlone and Lough Ree; in Clonmacnoise he noted inscribed stones that no one could read. At every stage he recorded place-names and popular customs: 'After supper I would get an old tale in Irish, and after all the family had gone to their rest one of the servants would begin to tell an old tale in bed without a respite until he lulls everyone in the house to sleep' (translated). Jones proceeded to Roscommon, to Ballinrobe, where he saw, running parallel with Lough Mask for four or five miles, a 'Rock of limestone with many strange lines' (translated) like those near Pontsticyll and Ystrad Fellte in Breconshire. He then went past Lough Corrib, where floods prevented him from viewing its limestone pavements and karst features, to Balatraw, the Maumturk mountains, and to Croagh Patrick, where he was unimpressed by the flora. He commented on local folk customs and provided early evidence of the importance of the potato:

> They [the people] weave cloth and flannel and generally knit stockings, but spin their linen with a hand wheel with the distaff under the belt like the Gwent pitcher women and those who studiously try to weave crewel in our country while singing a *cwndyd* [song or carol]. They enjoy poetry very much sitting on a bundle of straw in the middle of the floor foot to foot warbling and singing their Irish: as far as I saw, all the work of the people except

at harvest time is scorching their shins as they roast potatoes, or going from one neighbour's house to another to recite old tales and take tobacco. (translated)

He was disappointed not to find anything noteworthy on Achill Island but found gravel pits near Killala interesting. On 6 February, having learnt that Lhwyd planned to meet his assistants in Connaught, Jones warned him that he should expect to sleep on the ground on straw beds and offered him advice about his route:

If you design for ye County of Donnegall, Sligo is ye next in Conacht where you shall find good sport, viz. from Sligo Town to KnockRi [Knocknarea] where Crystalls are found, and then along ye shoar to Ballisydâr [Ballysadare], there are some gravel pits in those parts but abundance of limestones and from Ballisydar to Cas. Conacht and Kilala abt wch I think is \a good/ place for Lithosgoping; keeping ye seashoar & cutting indentures to ye gravel pits wch are not so numerous as abt Kilala, you may find heuraka [discoveries] from Sligo to Kilala 30 miles dist: ye barony of Erras [Erris] for iron oar and amber grease \the par. of/ Crossmelin [Crossmolina] & \about/ Lough Conn 6 miles distant from Kilala are places very good for Lithoscoping \about/ Lough Carow & Lough Mesk & Bellenrope are also very good. Bonevin [Nephin?] mam camp & mam trasna, mygerlin diewl [Magairli an Deamhain?] are high & rocky mountains for plants.'[49]

Lhwyd and Parry returned to Ireland from Campbeltown towards the end of January 1700. On 29 January Lhwyd wrote to Thomas Molyneux from Ballymoney, near Coleraine, to tell him that he intended to journey anticlockwise around the west of Ireland; after spending two or three weeks in Ballymoney he would proceed to Donegal and then to Connaught and Kerry. Since William Jones and Wynne had enjoyed 'such good success in their Several Pilgrimages' the previous year, Lhwyd had decided to remain in Ireland until midsummer to 'review the same places'.[50] Campbell has plotted Lhwyd's journey to Lough Larne, Castle

Doe (February), then on by 12 March to Sligo, where he remained until May, using it as a centre for expeditions further afield.[51] Lhwyd travelled to Longford in early May to meet 'one Teague o Roddy [Tadhg Ó Rodhaige] (*c*.1614–1706) ... an Excellent Irish Antiquary' who 'promis'd his Solution of some Queries I left with him',[52] and then to Roscommon, Clonmacnoise, Galway and probably around Lough Corrib and Lough Mask, venturing perhaps to Croagh Patrick and finally to Westport. By 8 July Lhwyd had returned to Galway, whence he sent Molyneux some plants and listed his discoveries; he intended to be in Cork 'about a Fortnight hence', and embark there for the 'West of England'.[53] By 22 July he was in Killarney, where he purchased a manuscript from Têg Moynihan[54] and then made his way to Cork.

When and where the two teams reunited is unknown; perhaps it was then that Jones drew the group of stone circles which lay 'Within half a Mile of Kyng [Cong] in a field on the Right Side of the Road as you go to Ballinrope'.[55] Other illustrations in BL, MS Stowe 1024 of Irish megalithic sites include tombs at Wardhouse (Co. Leitrim), Ardnaglass Upper (Co. Sligo), Ballina (Co. Mayo) and Magheracar (Co. Donegal).[56]

Although Richard Heaton (*fl.* 1602–60) made a number of discoveries in the 1630s,[57] the systematic study of Irish plants did not commence until the mid-seventeenth century.[58] Work then proceeded apace, so that by 1699 Ray could warn Lhwyd: 'I doubt whether *Ireland* will answer expectation, because it hath been already searched by skilfull & industrious Herbarists as Dr. Sherard & others.'[59] While tutor with the family of Sir Arthur Rawdon (1660–95) at Moira, Co. Down, in the early 1690s, William Sherard (1659–1728) botanised extensively and contributed several new Irish plants to the 1696 *Synopsis*.[60] He had worked mainly in the north-east of the country, but Lhwyd's route lay mainly in the west and south. Lhwyd began around Loch Larne, where (writing probably to Thomas Molyneux) he said that he had found 'plenty of a very rare plant mention'd by Mr Ray in his Synopsis but not well known to him or any other Botanist viz. Vaccinia rubra folijs myrtiis crispis' [*Vaccinium vitis-idaea*].[61] He told Lister mid-March that 'we have some hopes of adding a litle to Mr Ray's *Synops. of Plant*s, for we have already met with *Sanicula guttata, Sedum serratum folijs pediculis oblongis insidentibus* [*Saxifraga spathularis*], *Sanicula montana minor* ... and two or three more wch I doubt of, not having seen their Flower or seed' (G, p. 427). 'Sanicula guttata' is *Saxifraga hirsuta*, one of Lhwyd's most important Irish discoveries. He proceeded through Donegal and Sligo to 'the Mountains of Ben Bulben and Ben Buishgen [where] we met with a number of the rare Mountain Plants of England and Wales, and three or four not yet discover'd in Britain' (G, p. 432). In 'the Moors of the County of Mayo \& Galloway/' Lhwyd found 'a very elegant sort of Heath so common that the people have given it the name of Frŷch Dabeôg i.e. Erica (Sancti) Dabeoci & sometimes the women carry \ sprigs of/ it about them as a Preservative against Incontinency.'[62] This was *Daboecia cantabrica*, St Dabeoc's heath, first gathered by Tournefort in northern Spain before 1694.[63] On one of the Aran Islands (Campbell suggests Aranmore),[64] Lhwyd found

> great plenty of the *Adianthum verum* [*Adiantum capillus-veneris*] and a sort of matted Campion with a white Flower, which I bewail the loss of; for an imperfect sprig of it was onely brought me and I waited afterwards in rain almost a whole week for fair weather, to have gone in Quest of it. (G, p. 432)

The *Adiantum* was a plant familiar to Lhwyd since he had been the first to discover it, probably in 1697, in Glamorgan, 'growing very plentifully out of a marly Incrustation' at Barry Island and Porthceri.[65] In the far south-west, the mountains of Kerry also provided a number of rare plants. Lhwyd recorded finding *Alchemilla alpina* on 'the mountains of Dingle'.[66] Since Brandon Mountain, far down the Dingle peninsula, is one of the very few sites in Kerry where *Saussurea alpina* is found, it is possible that he had ventured so far despite the Tories, who 'obliged us to quit those \Mountains much/ sooner than we intended' (G, p. 457). Many of his specimens were sent to Oxford for further study, others to Ray, to Sherard, who identified several of them,[67] and to Richardson. The loss of most of Lhwyd's papers means that what would have been a major contribution to Irish botany[68] can be only partially reconstructed from his correspondence and from the extensive notes in his copy of Ray's *Synopsis*.

Lhwyd found many new fossils in Ireland. In 1709, when sending Irish fossils and plants to Samuel Molyneux (1689–1728), Lhwyd said that he knew of 'no place in the three Kingdoms that have better Variety of uncommon figured Fossils' than 'Ben vuisgen & Ben Bulben'.[69] Modern geologists agree with Lhwyd's assessment of the importance of the area.[70] Since finding these 'curiosities' had cost Lhwyd 'a great deal of Expence of Time & Money', he requested Molyneux not to communicate them to 'Woodward or Pettiver or any other of Our London Virtuosi, who without any Labour or Expence may publish what has cost others both'. Woodward was an old enemy, and Petiver had offended Lhwyd by pillaging a cargo of Maryland specimens sent to him by Hugh Jones (G, p. 462).

A major objective of Lhwyd's tour of Ireland and Scotland was finding old Gaelic manuscripts. As shown by the fate of the Ó Gnímh manuscripts, by this time the Gaelic/Irish manuscript tradition was in serious decline and manuscripts were at risk of loss or casual destruction. Fortunately, as Lhwyd told Robinson in August 1700, he had 'in divers Parts of the Kingdom picked up about 20 or 30 Irish Manuscripts on Parchment; but the ignorance of their Critics is such, that tho' I consulted the chiefest of them, as Flaherty … and several others, they could scarce interpret one page' (G, p. 431). Lhwyd met Roderick O'Flaherty (*c*.1629–1716?), the impoverished author of *Ogygia* (London, 1685), at his house, Parke, near Galway, in July 1700. They soon began a correspondence which was of value to Lhwyd when he started work on Irish-language material for the first volume of the *Archæologia*.[71] Lhwyd attributed the shortcomings of Irish scholars to the 'want of a Dictionary, which it seems none of their Nation ever took the trouble to compose. I was informed (but how truly I know not) they have lately printed one at the Irish College in Lovain' (G, p. 431). This was Mícheál Ó Cléirigh's *Foclóir nó Sanasan Nua*, a monolingual Irish dictionary published in Louvain in 1643. As it soon became a very rare book, Irish scholars made manuscript copies of it. Lhwyd was shown such a transcript by the scribe and collector Arthur Brownlow (1645–1711),[72] hence his uncertainty whether the work had been published. Considine considers Lhwyd's dismissive comments on Irish scholars to be a better indication of his own ignorance than theirs and emphasises that as late as 1700 Lhwyd was still unaware of the medieval Irish lexicographic tradition.[73]

Lhwyd (over)confidently told Robinson that, given a dictionary, 'I should not despair of being in a short time able myself to understand' the manuscripts he had acquired:

> Tho' many of them being but insignificant Romances it would scarce quit the pains. What I most value amongst them are their old Laws, which might give some Light to the curious as to many of their National Customes; and some of their old Poems; but all are of use to any that would compose a Dictionary of their Language; which was anciently ... doubtless very copious. (G, p. 431)

He was even more dismissive of the value of most of his manuscripts, telling Richardson: 'They contein litle of Authentic History excepting what is insignificant, their Genealogies.' However, 'several Critical uses may be made of most old Rubbishes.'[74] Later scholars would agree with his view of the importance of the law manuscripts, authoritative texts which constitute the largest part of the existing corpus of Old Irish law. What Lhwyd had dismissed as 'insignificant Romances' were major, sometimes unique, collections of early Irish literature. These included the Yellow Book of Lecan, a major source of the tales of the Ulster Cycle, including early versions of *Táin Bó Cúailnge* and associated tales, and the twelfth-century Book of Leinster with its rich collections of tales, *Lebor Gabála Érenn*, genealogies and verse. Lhwyd had also acquired much grammatical and lexicographic material, some ancient like *Auraicept na n-Éces*, others more recent. After his death his Irish manuscripts were acquired by the Sebright family, and in 1786 Edmund Burke persuaded Sir John Sebright (1725–94) to donate to Trinity College Dublin over forty Irish manuscripts, a collection that has been described as one of 'the most notable surviving witnesses to Lhwyd's greatness'.[75]

Isle of Man

It seems that Lhwyd never visited the Isle of Man,[76] nor does it appear in the places listed on the title page of the *Archæologia*. A few Manx words, cited in the appendix to the British Etymologicon, pp. 290–8, the first to appear in print,[77] were drawn from a Manx vocabulary, *Geirieu Manaweg*, which survives in a fair copy prepared by William Jones.[78] While in Dublin in early 1700 Jones was apparently translating the first section, the 'North-country words', of Ray's *Collection of English Words* into Manx and Irish.[79]

Cornwall, August 1700–January 1701

Through his friendship with Aubrey and Wylde (see Chapter 4), Lhwyd had been interested in Cornwall and Brittany since the early 1690s. In February 1692 Aubrey told him about 'an ingeniose young Cornish Gent who hath been in Catalonia & he finds there and about the Pyrenean hills severall \many/ *British* words'.[80] Lhwyd was sceptical:

> I did not think that any young Cornish gentlemen had understood British ... 'tis possible the gentleman having picked out five or six parallel words (which is easily done of any language in these parts of Europe) took it for granted from their guttural pronunciation, that there might be many more. (G, p. 157)

In his 1692 letter Aubrey put Lhwyd on the track of Cornish scholars and manuscripts, since the 'Cornish Gent' had told him

> that one Mr [John] Keygwin (1642–1716)[81] ... is a great master of the Cornish tongue: he hath the greater part of the Bible in Cornish MS and also the History of the Passion of our Saviour, entituled Mount Calvary, in Cornish verse (in the old Saxon character as also in the Bible) ... and he hath other MSS of which I shall have a Catalogue ... [my friend] will putt him upon making a Cornish Dictionary.

Lhwyd and his companions sailed to Cornwall in August 1700.[82] Since his letter to Robinson from Penzance dated 25 August says nothing about Cornwall (G, pp. 431–3), he had probably just arrived. Lhwyd set about gaining letters of introduction. A key figure was Sir Jonathan Trelawny (1650–1721), bishop of Exeter.[83] Although an ambitious politician rather than a serious scholar, Trelawny had commissioned translations of Cornish texts from John Keigwin. Through his secretary, Thomas Newey (1658–1723), a canon of Exeter, Trelawny replied on 3 September 1700 to Lhwyd's request for assistance, offering to show him his 'Cornish books' and suggesting that Sir Joseph Tredenham (*c.*1641–1707) and Lewis Stephens (1654–1725), vicar of Menheniot, near Liskeard, would be useful contacts. Stephens, an expert on seaweeds, had contributed to the second edition of Ray's *Synopsis*, as had his friend Walter Moyle.[84] Stephens himself does not appear to have been particularly interested in Cornish but acted as a link between Lhwyd and local Cornish enthusiasts. Though resident in London, Stephens's nephew, John Anstis (1669–1744), a herald and antiquarian, introduced Lhwyd to John Moore (b. 1648), rector of Helston, who referred to Cornish as 'my own Mother Tongue'.[85] Moore, in his turn, introduced Lhwyd to the antiquary and collector Richard Erisey of Mullion (d. 1722), a descendant of one of Drake's privateering captains.[86]

Lhwyd sent Robinson from Penzance 22 September a brief letter about Cornish marine life and plants.[87] Its brevity indicates that his linguistic and antiquarian interests now predominated. Lhwyd soon realised that he should have searched for Cornish minerals and fossils. In 1703 Lhwyd asked Thomas Tonkin (1678–1741/2) to 'Collect & procure all the Variety you can hear of, of the tin-oars: for tho' I thought I was Tolerably well furnished, Yet I find by the Swede [Angerstein], who was last Winter in your Country, that I have but a poor Collection.' Lhwyd also remarked that they had 'met with no Fossil Shells, or other Marine Bodies in Cornwall' (G, p. 487). Although much of Cornwall is made up of unfossiliferous rocks, fossils can be found in certain places, particularly along the north coast between Bude and Boscastle.[88] In late September or early October, Jones took advantage of an exceptionally low tide to collect starfish

and described in detail another creature 'that they call pôl men' (translated).[89] Drawings in BL, Stowe MS 1023 include a depiction of
'Krach y Mor Meirionens prope Straton', not a barnacle, as Jones
thought, but part of a piddock shell.[90]

Francis Paynter of Boskenna, who had met Lhwyd in Penzance on
23 September, subsequently invited him and his assistants to dine with
him on the 29th to meet John Hickes (1658/9–1734) of Trevithick,
who wished to offer his help.[91] Lhwyd met Tonkin at Lambrigan on
27 August, when the latter gave him a letter of recommendation to
'Mr. Chancellor Pennick',[92] and lent Lhwyd Carew's *Survey of Cornwall*
(1602) and some other items. When William Jones returned these,
Lhwyd's covering letter dated 15 October asked Tonkin to provide Jones
with letters of introduction because

> where we have no acquaintance we find the people more suspicious
> and jealous (notwithstanding we have my Lord Bishop's approba-
> tion of the undertaking) than in any country we have travelled.
> And upon that account I beg the trouble of you, when he [Jones]
> leaves your neighbourhood, to give him two or three letters to any
> of your acquaintance more eastward. Mr Pennick not being at
> home, we have been strictly examined in several places; and I am
> told the people, notwithstanding our long continuance here, have
> not yet removed their jealousy. (G, p. 434)

Tonkin subsequently defended his countrymen:

> Mr. *Lhuyd came* into the country at a time, when all the people
> were under a sort of panick, and in terrible apprehension of thieves
> and house-breakers; and travelling with his three companions
> (with knap-sacks on their shoulders) on foot … prying into every
> hole and corner, raised a strange jealousy in people already so much
> alarmed: though this alarm … was without the least foundation,
> and at last discovered to be the contrivance of some designing
> neighbours, to get money for their assistance in this pretended
> danger … At *Helston*, as Mr *Lhuyd* was poring up and down, and

making many enquiries about Gentlemen's seats, &c. he (with his companions) was taken up for a thief, and carried before a Justice of the Peace; who, on opening my Letter was very much ashamed at it, and treated him very handsomely.[93]

On 15 October Lhwyd had outlined to Tonkin his plans for the tour: 'now it so happens, that I [with Parry] take the south coast, and leave the north to the bearer [Jones, with Wynne], to copy such old inscriptions as shall occur, and to take what account he can of the geography of the parishes.' Lhwyd's own itinerary can be partially reconstructed from his correspondence and from his twenty-three parish descriptions preserved in Bodleian, MS Rawlinson D 997.[94] Notes made by the other group have not survived, but the subjects of drawings of earthworks, stone circles, inscriptions and other antiquities preserved in BL, Stowe MS 1023 are located in another twenty-one parishes. Although they may be later copies of drawings by William Jones, these indicate some of the parishes he visited. Lhwyd did not adhere strictly to the plan he had outlined to Tonkin, for his surviving parish descriptions are mainly of inland parishes, and Jones's drawings are not restricted to the south coast.[95] The two parties may have met at certain places; Lhwyd noted a house near St Colum, and Jones's drawing of the Nine Maidens stone row near St Colum Major could have been part of the same journey. Following his return, Lhwyd told Mostyn that they had 'rambld through almost all the parishes of Cornwal' (G, p. 443). Notes made by Lhwyd and his assistants are extant for forty-five Cornish parishes of the 161 listed in *Britannia*. Unlike the Welsh returns published in *Parochialia*, the Cornish parish descriptions, often in a mixture of English and Welsh, remain unpublished, apart from Cambourne and Illogan, which are printed in Williams.[96]

Once back in Oxford Lhwyd sent Henry Rowlands an account of his tour (G, pp. 439–42). After describing his misadventures in Brittany, he noted that while Cornish was 'confined to half a score parishes towards the Land's End', Breton was 'the common language of a country almost as large as Wales'. Pride of place went to his major achievement in Cornwall:

I have procured transcripts of the only three manuscripts extant in the Cornish. The oldest is a Poem of the Passion of our Saviour, written on parchment about two hundred years since; the others contain several Opera's or Plays, all out of the Scripture.

After noting he had discovered 'some old inscriptions not observ'd before' he remarked that

The modern Cornish seem to me a colony of the Armoricans from their language and habit … for one may observe from the names of places that another people once possess'd that country, as one may from the names of places in some parts of Wales gather that the Irish Nation once inhabited there particularly in Brecnock-shire and Caermarthen-shire, where the Lakes are call'd Lhychæ and the high mountains, Bannæ; as they commonly are throughout the highlands of Scotland and Ireland. (G, p. 442)

Lhwyd adhered to the end of his life to this inverted theory of the relationship between Cornwall and Armorica. After noting 'towards the Land's End' amulets which the Cornish called 'a Melprev or Milprev' he observed that

Cornwall affords store of those barbarous monuments we have in Wales … viz. Meini Gwir (or stones pitch'd circularly) Kromleich, Kryg or Gorsedh, Kaer, Karn, &c. Of these … in Bretagne, we met with only the Kryg and Caer, but were inform'd also of the circular stones.

A few weeks later he sent Richard Mostyn a detailed account of his visit to Brittany but added little about Cornwall except that

The onely four Cornish Books remaining were communicated to me, besides many other Favours by ye Bishop of Excester, and I have copies of each of them. That Countrey affoarded some ancient Inscriptions like those added to Camden in Wales: & both there, and in Irland and Scotland *Caer, Carn, Din,* & *Cromlech* are frequent and often (allowing for pronunciation) distinguished by the same names. (G, p. 445)

The apparent discrepancy in Lhwyd's two accounts may be because he encountered three texts, but four 'books'. The 'original' manuscripts of the *Ordinalia* and *Gwreans an bys* were in Oxford, and the 'Passion poem' text had been transcribed. There may have been more than one copy of the text available to Lhwyd and this may be the fourth book to which he refers. Setting out his future plans to Tonkin 8 February 1703, he described the two Cornish manuscripts in Oxford that Tonkin wished to have transcribed, closing with the words:

> Those few things that occurred to me in Cornwall, which are chiefly Inscriptions, and a Vocabulary as Copious as I can make it, I design to insert (God willing) in my *Archæologia Britannica*; which I hope to Print some time this next Summer. (G, p. 484)

As well as examining and transcribing manuscripts Lhwyd also gathered rhymes and poems, some from written sources, but others apparently collected orally. He compiled glossaries and word lists based on his reading and conversations,[97] in preparation for his planned comprehensive dictionary. At the end of his Cornish glossary in MS Ashmole 1820a, he requested readers to send him material:

> Any old Proverb or Rhythm would be very acceptable such as,

> > *An Lavar gôth yw Lavar gwîr*
> > *Ne vedn navraldoz vâz in Tavaz rehîr:*
> > *Bez dên heb davaz a gollas e dîr.*

Lhwyd considered this verse to be particularly important, quoting it with minor variations, some four times. He explained to Tonkin in March 1703 that

This sort of verse was, for what I can yet find, the oldest, if not the only verse amongst the ancient Britons: for 'tis the oldest in our Welsh books, and I have heard an old fellow repeat one of them in the highlands of Scotland; and had another from the Clerk of St. Just.

He then quoted the verse with an English translation:

> The old saying is a true saying,
> A tongue too long never did good;
> But he that had no tongue, lost his land. (G. p. 485)

He was to develop this idea in the *Glossography*, pp. 250–1:

The sort of Verse I find most common amongst our oldest Remains, is that called *Englyn Milur* in *Jo. Dav. Rhŷs's* Grammar p. 184. And in regard I have (tho' but rarely) heard the same in the Shire of *Argile* in *Scotland*, and also in *Cornwal*, I am apt to conclude it one of the most Ancient, if not the very oldest sort of Verse we ever had; and that 'twas in this sort of Meeter the *Druids* taught their Disciples … That this is Ancient enough to have been the Verse used by the *Druids*, is Manifest from there being some Traditional Remains of it at this day, in *Wales*, *Cornwal* and *Scotland*; tho' it be immemorial when any such were last made. And that it really was used by them seems also highly probable in that a great number of the *Welsh Englyns* of this sort have always some Doctrine, Divine or Moral, in the Conclusion; the rest being often insignificant and serving only as Meeter thereunto.

He cited as examples Welsh gnomic verses and 'Elegies and Historical Memoirs', such as *Englynion y Beddau* (Stanzas of the Graves), closing with, 'What I happen'd to hear in Cornwall was only this one *Englyn, An lavar kôth yu lavar guir.*'

Earlier in the *Glossography* (pp. 239–40), Lhwyd had argued that whatever the present differences, the early forms of Welsh, Cornish and

Breton were largely indistinguishable from one another: 'If there be any writings extant therein, so old as the 7th, 8th, 9th, or 10th Centuries; I presume they are scarce distinguishable as to Grammar from Welsh'. Having failed to 'meet with no very ancient writing that seem'd to be Cornish', Lhwyd tentatively suggested that a sequence of three-line *englynion* in the Red Book of Hergest, an 'Elegy on the Death of Gereint ab Erbyn, a Nobleman of Cornwal or Devon about the year 540' might be an example of ancient Cornish. The poem is actually in Middle Welsh. When Lhwyd composed an elegy in Cornish on the death of William III in 1702, he used this metre, claiming it to be '*Ad normam Poetarum seculi sexti*' (According to the rule of the poets of the sixth century), as he also did in his Breton elegy to Prince George of Denmark in 1708, written, he said, '*ad Bardorum veterum exemplar*' (according to the model of the old bards).

It is not inherently improbable that Lhwyd should have been given a Cornish verse by someone in St Just, a Cornish-speaking parish. Were it a genuine relic of Middle Cornish, a surviving fragment of a lost literary tradition, its importance would obviously be very great. Sceptics have argued that such a verse could not have survived so well over a period of linguistic and cultural decline, that there are 'modern' traits in its language and that the proverbs that are the content of the verse can be found in seventeenth-century English printed sources. The scholarly consensus is that it was a seventeenth-century composition.[98] Lhwyd never quoted the 'similar' verse from Argyll, and it is not to be found in his surviving notes. Although Lhwyd (or more probably) one of his assistants had fabricated a pseudo-archaic Welsh text, an 'Old Welsh' *englyn* inserted into a Margam Abbey charter,[99] it seems unlikely that Lhwyd composed the Cornish *englyn*. Since the Margam Abbey *englyn*, intended to tease 'experts', was not for general circulation, its intention was quite unlike the Cornish verse. Lhwyd's theory of a common Celtic culture may have predisposed him to accept the *englyn* as genuine.

Although Lhwyd's Cornish tour had not led to any significant discoveries in natural history, he had recorded monuments, inscriptions and popular beliefs. He had acquired first-hand experience of the language (enough to venture to write in Cornish), its literature and its current condition, and the manuscripts and other fragments that he had safeguarded enabled him to begin compiling Cornish glossaries. Cornish scholars had been ready enough to assist his researches but proved strangely reluctant to subscribe to the *Archæologia*. Lhwyd could find no more than four Cornish subscribers: Walter Moyle, Sir Joseph Tredenham, Bishop Trelawny and John Hickes. In March 1703 he appealed to Tonkin to subscribe, 'let the sum, before hand be as small as you please' (G, p. 486). Tonkin duly obliged and his name was included in the list in *Glossography*.

Brittany, January–March 1701

Lhwyd's interest in Brittany, like his interest in Cornwall, dated back to the early 1690s. In 1690 or early 1691 Edmund Wylde had introduced him to the *Originum Gallicarum Liber* of Boxhornius (Marcus Zuerius van Boxhorn (1612–53)). Although Lhwyd considered Boxhorn to be full of himself, '*sui plenus*', he found his arguments for the affinity of British and Gaulish 'very plausible'. Wylde had also spoken of a French scholar who was discussing the 'origin of the Galles' (G, p. 134). This was Paul-Yves Pezron (1639–1706), whose discussion of chronology, *L'Antiquité des Temps Rétablie* (1687), had led him to Gaulish and to his native Breton.[100] By 1693 Lhwyd had (wrongly) come to believe that Pezron had published a book on Gaulish origins under the title *Antiquitas Gaulois*. When he failed to find a copy, he sought to borrow one from Wylde (G, p. 176), but his hopes of seeing whatever Breton book it was that Wylde possessed (perhaps the 1499 *Catholicon*) were dashed by Aubrey, who told him in April 1693 that Wylde 'will lend no booke to anyone'.[101] Lhwyd had mentioned to Lister in 1696 his intention 'if we have peace' to visit 'Britaigne' (G, p. 314). The signing of the Treaty of Ryswick in the autumn of 1697 brought a temporary suspension of hostilities with France and made such a visit possible. Lister took advantage of this to visit Paris, where he bought

the Works of *Pere Pezron* … a very Learned and very disinterested Author … He is now upon giving us the *Origin of Nations*, where he will shew, that *Greek* and *Latin* too, came from the *Celtique* or *Bas-breton*; of which Country he is. He told me he had 800 *Greek* Words perfect *Celtique*. I settled a Correspondence betwixt him and Mr. *Ed. Floid*; which he most readily granted, and which he said he had long coveted.[102]

For some reason Pezron failed to reply to any of the letters Lhwyd sent him between 1699 and 1701 (G, p. 441). By 1698 Lhwyd had gained a better idea of Pezron's views (perhaps from Lister) and expressed his reservations:

> [Pezron's] Notion of the Greek, Roman & Celtic Languages being of one common origin agrees exactly with my observations. But I have not advanced so far, as to discover the Celtic to be the Mother-Tongue, thô perhaps he may not want good grounds, (\at least/ plausible arguments), for such an Assertion. The Irish comes in with us, & is a Dialect of the \old/ Latin, as the British is of the Greek. But the Gothic or Teutonic, thô it has also much affinity with us, must needs make a Band apart. (G, p. 400, date corrected from *EMLO*)

Since Lhwyd was now so near to Brittany, Lister and Robinson urged him to venture there, the latter suggesting he might subsequently visit Paris.[103] Lhwyd told Richardson on 21 October that he had

> at last finishd those Rambles I proposd in the printed Paper ... But finding myself here, within twenty hours sail of Armorica; & presuming what few Observations I have made, may be somwhat better cleard by a journey into that Countrey, I am resolvd ... to step over from one of these ports & to return through Paris.[104]

Lhwyd considered it prudent to request the consent of the Vice-Chancellor before venturing overseas and was told at the end of October that there was no objection to his going abroad.[105] Jones and Wynne were sent back to Oxford on 13 November 1700, leaving Lhwyd to make arrangements to visit Brittany, mainly through Tonkin, whose father-in-law, James Kemp of Penryn, traded in Morlaix (G, p. 435). By the end of November preparations for a visit of two or three months were well advanced. Lhwyd hoped that letters to Breton antiquarians and clergy, explaining the scholarly purpose of his enquiries, would avert the suspicions he had aroused in Cornwall. While awaiting a favourable wind, Lhwyd told Richardson that after staying no more than a month in Paris he would 'consult my pocket &c whether to return immediatly for Oxford through London or through Flanders and Holland. My utmost ambition would be to see the Pictures of the Carps, Eels, Pearch in the Slate pits of Isleb [Eisleben] in Germany.'[106] Lhwyd and Parry sailed from Falmouth, probably on the *Marie*, and arrived at St Malo on 16 January 1701.[107] After hearing that several English merchants had been arrested as spies, Lhwyd obtained additional letters of recommendation from merchants and clergy in St Malo and Morlaix and hoped to avoid attention in Saint-Pol-de-Léon, almost forty miles to the northeast of Brest. As Lhwyd subsequently told Tonkin, 'the Intendant [des Marines] of *Brest* ... having a little before received a check from Court for some negligence was pleased, by way of making amends, to exercise his double diligence on me.'[108] Within some three weeks Lhwyd had been found, as he told Henry Rowlands, 'busy in adding the Armoric words to Mr Rays *Dictionariolum Trilingue*, with a great many letters and small manuscripts about the table'. Lhwyd and Parry were detained and their books and papers 'ty'd up in a napkin', sealed and confiscated. Although a senior clergyman offered to stand bail, they were taken to Brest castle as possible spies. A letter from the Theologal of the cathedral of St Pol-de-Léon spared them from joining the English merchants in the town jail, but initially they were not permitted the usual allowance to purchase food and drink. Since they were confined in a ground-floor room, Irish soldiers in the garrison handed them rations through the window. They were then allowed fifteen pence a day 'with tolerable good white-wine for three pence a quart'. Lhwyd drew up an elegant Latin petition stating his case and indicating his testimonials (G, p. 438). Although the Intendant claimed that he 'was not conversant in that language', the names of two influential Parisian abbés (Abbé Philippe

Drouin and Abbé Guillaume Du Bois) mentioned in the petition may well have convinced him that Lhwyd (unlike the English merchants he had earlier detained) was too well connected to be trifled with. The following day Lhwyd had to formally identify his papers, which were then handed to an interpreter. He could make nothing of documents in Welsh and Cornish and, to avoid displaying his ignorance, after some nine days told the authorities that none of Lhwyd's papers 'related to State-matters ... upon which we had all our papers restored ... and [were] order'd to depart the kingdom' (G, pp. 439–40). Lhwyd was disappointed that 'our coarse welcome in France' had

> prevented almost all the enquiries I design'd into the language, customs and monuments of that province. For all we could do was but to pick up about twenty small printed books in their language which are all (as well as ours) books of devotion, with two folios publish'd in French; the one containing the History of Bretagne, the other the Lives of the Armoric Saints. (G, pp. 440–1)

One of the books that he acquired was *Le Sacré Collège de Jésus ... ou l'on enseigne en armorique ...* (1659) by Julien Maunoir (1606–83); on the title page Lhwyd wrote, rather ruefully, in Welsh: 'I bought this book too expensively in Saint Paul de Leon in Brittany in 1700 [*sic*]. It was not on sale anywhere, so I left in its place Dr Davies's Dictionary which is a hundred times its value in the monastery of the friars minor' (translated).[109] He had not been able to see, let alone acquire, the 1499 *Catholicon*, which he described to Humphrey Foulkes in 1708 as a 'Dictionary I have been long in quest off'.[110] His imprisonment and expulsion had prevented Lhwyd from having 'time to consult either men or books, or to view any of their old monuments, so that I shall be able to say little of that country, besides what relates to their language' (G, p. 441). Writing to Mostyn towards the end of April, Lhwyd added that he had 'found the monks evrywhere obliging enough, but 'twas not my Fortune to find any amongst them anxious in my studies ... They could tel me nothing of Father Pezron' (G, p. 445). Lhwyd considered that Breton is much ye same with the Cornish; & both so near ye Dialect of South Wales: that in a months time at farthest a Welshman may understand their writings; but as to the speaking part their affinity creates some confusion. 'Tis spoken at least for a Hundred miles, and their Gentry & Merchants speak it in their Great Towns; but more corruptly than ours in N. Wales, and they seem to have been more discourag'd by ye Mounsieurs jeering them than those of sense & Education are amongst us. (G, p. 444)

As well as making these comparative observations, Lhwyd believed he had learnt enough to allow him to describe the phonology of the language and its dialects and to attempt to compile the dictionary that the language required.[111]

THE GLOSSOGRAPHY
AND AFTER

On returning to Oxford in the first week of March 1701 'after a tedious Ramble of four years', during which he had covered over three thousand miles, Lhwyd said that he was 'setl'd (if it please God) for the remainder of my time'. His first task, which took precedence over 'culling out the pertinent part of my collection and digesting it for the Presse' (G, p. 443), was to restore order at the Museum. Before embarking on his travels, Lhwyd had appointed William Williams as Underkeeper and deputy. He quite unrealistically expected Williams, with the assistance of the Under-Librarian, Robert Thomas, to run the Museum during his absence. Both of them were poorly paid and lacked the status to enforce payment of users' fees.[1] Indeed, poor Thomas was 'soundly beaten for his impertinence' in demanding payment from a 'youthfull Fellow of New Colledge with ... a Gentleman of his acquaintance'.[2] As he became increasingly disgruntled, Williams absented himself for lengthy periods. To make things worse, he and Thomas did not get on, and the latter resigned in June 1699. A stop-gap appointment ended when 'one Jones who was then Librarian' left, and in May 1700 the Visitors unanimously appointed a new Under-Librarian and *de facto* deputy, Richard Massey (1670–1743).[3] According to Lhwyd's friend John Ellis (1674–1735), Massey discharged 'his trust with great care & diligence', bearing 'the whole trouble of the Musæum, as for Mr Williams he made no other than a Sinecure of it.'[4] Massey had to leave Oxford in early November 1700 'because of some urgent buisness that calls me into

Cheshire',[5] leaving the Museum virtually unattended for a week or so. According to Ovenell, Williams had resigned in September 1700,[6] but Massey expected his imminent return in November. Williams subsequently remained in contact with Lhwyd and his assistants until he was drowned in a shipwreck off Glamorgan in August 1701.[7]

When William Jones and Robert Wynne reached Oxford on 20 November they were appalled by the state of the Museum and wrote to Lhwyd four days later. According to Jones:

> everything here is in disorder and I don't yet know what is missing. Some of the pictures that should be on the staircase have been moved to your room below, and the labels that used to belong to them lost. I looked over the best cabinet in which about twenty-four small objects are missing ... Mr Williams is expected here within nine days who I hope will be able to give a full account of everything.[8] (translated)

The books were untidy; they had found the 'middle study' open, and could not tell whether everything was as Lhwyd had left it. Wynne had now locked the study; the bolt of the lock had been forced, since the key was in one of the twenty-three big boxes sent from Dublin. Twenty-two of these had arrived and one more was expected. Wynne confirmed Jones's account; 'tis more than we can do, to express wt disorder we find here'. It grieved him to see 'the lower door of the middle study wide-open', and he added that Jones had the previous day 'dismisd ye young man [a batteler called Williams] we found in trust, & tomorrow I will take him to task.'[9] After noting moneys collected and payments made, Jones assured Lhwyd that 'Whatever is now out of its order and place in the Museum I myself shall see that it is restored to its place as well as I can and I shall take so much care with everything that I hope no one can find fault with your faithful servant Will. Jones' (translated).[10]

Lhwyd set about doing what he could, but the damage would take years to repair. In early June he explained to Richardson why he had not written sooner: 'I have been in such a Hurry most of my time since my

Return, partly in restoring the Things of the Museum \(which I found in some confusion)/ to their old places, against our Visitation, which is yearly on Trinity Monday [16 June]' (G, p. 457). During Lhwyd's absence there was a high turnover in the body of Visitors and Visitors were frequently represented by deputies.[11] The annual Visitations appear to have been superficial; as Williams told Lhwyd in mid-July 1698, the Visitors 'were inquisitive after yr return but found no fault in the Museum (such was either their carelessness or their good humour)'.[12] As well as attempting to restore order Lhwyd had to 'methodize' his new collection of fossils from Wales and Ireland, which was, as he told Richardson, 'almost equal to that which is Printed [in the *Ichnographia*]; but they chiefly excel in Coal-Plants, *Lithostrotia*, and *Modioli* of Fossil-stars' (G, p. 457).

The University recognised Lhwyd's achievements shortly after his return by granting him an honorary MA degree on 21 July 1701 on the grounds that although he was of almost twenty years' standing in the University, his prolonged absence 'to perfect himselfe in the knowledge of Natural History and Antiquities' had prevented him from proceeding to his degree.[13] The award resolved the long-standing anomaly that the head of a major University institution was not a graduate. In return, Lhwyd was required to deliver six annual lectures on natural history. His topic, the nature and origin of starfish and their fossilised remains, had long interested him and had been the subject of 'epistles' to Nicolson and to Archer in the *Ichnographia*. In 1702 Robinson encouraged him to write up the lecture for inclusion in a new edition of the *Ichnographia*.[14] Lhwyd eventually accepted this advice, placing an announcement in the *London Gazette*, 10–13 January 1708, which stated his intention to publish a 'second Impression' with 'An Appendix de Stellis Marinis'. Lhwyd's death put an end to the project, but amongst his papers were notes and drafts entitled '*de stellis marinis*'.[15]

Lhwyd's main unfinished business was the acquisition of Cole's collection. In May 1695 Lhwyd told John Lloyd:

> We are like to receive either this Summer, or the next, a very noble
> accession in the Museum; for Mr Cole of Bristow, who … has a
> Collection of Natural Bodies well near as considerable as that we

now have, has offer'd to give them [to] the University, on condition they'll print a Book of his composing, which contains some
observations he has made in Natural History: And they are very
like to comply with his Request. (G, p. 271)

Lister urged Lhwyd to 'close with C. if you can make sure of him; and
will doe best to visite him; being a man of substance, and much in the
circumstances of mr Ashmole, having noe childern'.[16] Lister's cynical
advice indicates that he, like Lhwyd, had realised that Cole might well
provide the Museum with an endowment to support the maintenance
of his collection. Cole, for his part, hoped Lhwyd would persuade
the University to publish his book. This argued that 'fossil shells have
chang'd their Species by long lying in the ground', a theory which was
unacceptable to Lhwyd, who told Lister in June, 'I fear I must dissem-
ble abominably before we can come to an agreement.'[17] In July Lhwyd
hoped to travel to Cole's country house in Wiltshire 'to treat with him
about the Museum' (G, p. 278), but Cole had to cancel the meeting.[18]
By early October Cole was disappointed to hear a rumour that 'the Vice
chancellor is grown cold as to my proposalls.'[19] After deprecating Cole's
jealousy of Ashmole, Lister commented: 'What can all this bosting be
about ... possiblie his figured stones ar many, but ... will he give land
or a rent charge upon it to endow the Musæum to make it worth the
while to looke after it, & to purchase yearlie something of Nature or
proper bookes to furnish it?' He added presciently, 'I am affraid he will
give you a great deale of trouble to noe purpos.'[20] Towards the beginning
of November 1695 the Vice-Chancellor sent Lhwyd to discuss terms.
By then Lhwyd had lost patience with Cole: 'the most vainglorious and
conceited man I ever met with, a fault which perhaps is chiefly owing
to his Education',[21] but persisted in his quest since, as he told Lister,
Cole had said that: 'the least Sallary he intended for the Keeper was
ten pounds a year; but added that perhaps 'twould be much more ... I
supose the most will be twenty, thô I presume he is very able to make it
fifty, being very rich, and having no children.'[22] Lister warned Lhwyd
that 'you have gott a Wolf by the eare, such a one as you neither can
well hold without monstrous slaverie, or well let goe, because you have

engaged your honour to him. much good may it doe you.'[23] Lhwyd responded that Cole

> perhaps ... designs an other Keeper for his Collection, but if so I'll break all measures; for as I brought him first to treat with the University, so I'll presume they'll dismisse him [the Keeper], when ever I advise them so to doe ... the main thing I humour him for, is the Sallary he intends for the Keeper. (G, p. 291)

Negotiations then had to be suspended since Lhwyd's expeditions for *Britannia* and then for the *Archæologia* took him away from Oxford, and Cole himself became involved in lawsuits. During Lhwyd's travels he received many friendly (and very lengthy) letters from Cole offering information about fossils along the banks of the Severn estuary and in south Wales, including 'Oyster Stones' sent by Lady Stradling 'out of the Rockie Cliffe joyning to her Garden at St Donnett'.[24] Cole also drew Lhwyd's attention to the fossils of Caldey and attempted to arrange accommodation for him in Tenby.[25] He gave Lhwyd specimens, including in March 1697 'as many of the Species of the Sea Starrs as I have in my collection'.[26] In return, Lhwyd sent Cole specimens and books, including the pointed gift of Redi's demolition of the theory of spontaneous generation of insects. Cole, a firm believer in the spontaneous generation of worms from horse-hair, was not convinced: 'It is an astonishing thing to me to see so many learned Men ... so much admiring [Redi] ... I will teach an ingenious boy of 10 yeares old substantially to confute him.'[27]

In early June 1701, Lhwyd told Richardson that 'Mr. Cole of Bristol, who these many years has promisd us his Museum, is very importunat with me to visit him at his Countrey House for a week or Fortnight, which is in Wiltshire, about a day's journey hence' (G, p. 457). After the visit Lhwyd expressed his doubts to Humphrey Foulkes:

> I have been most of my time in Wiltshire with an old Vulpone who has talked these seven years of contributing a very large colection of natural Rarities to ye Museum and of setling some small salary

upon't; but what he'll do time must shew. (G, p. 459, addressee corrected from *EMLO*)

Some two months later, on 30 August, Cole died intestate. Lhwyd redoubled his efforts, enlisting the help of his former colleague Obadiah Higgins, who spent three days viewing the collection in February 1702. Despite Lhwyd's pleas, the Vice-Chancellor vacillated and in July the collection was sold to Dr John Lane (d. 1740), a pioneer of smelting non-ferrous metals in the Swansea area, and subsequently vanished from sight. Lhwyd was very critical of the University, blaming in particular the Vice-Chancellor, who had 'demurr'd so long ... We might have had it scandalously cheap; for about £60 and ye Cabinets boxes and Glasses were worth half the money' (G, p. 469).

The loss of the salary Cole might have provided was particularly galling as Lhwyd calculated the costs of the tour and considered how to pay for the next stages of his project. The original intention had been to finance both the tour and the publication of the volumes by annual subscriptions. The initial response had been encouraging but, despite the efforts of Walter Thomas and others, not all subscribers maintained their payments and the sums received each year declined. After Lhwyd and Thomas had finally separated promises of support from subscriptions received, it appeared that about two hundred people had contributed almost four hundred pounds towards Lhwyd's researches between 1696 and 1700, some paying towards the cost of the tour and a copy of the first volume, others for the book alone. When it became clear that London booksellers would not publish the *Archæologia*, a book 'in a Subject so singular', Lhwyd had to find subscribers to finance its publication. He therefore issued a new prospectus, dated 10 July 1703, setting out the contents of *Archæologia Britannica, or An Account of the Ancientest Languages, Customs, and Monuments of the British Isles ... Tome I. Of The Languages*.[28] Towards the end of its closing paragraph, Lhwyd, with more honesty than tact, said that he was aware that the book would be of 'little Use' to many of the gentlemen who had contributed to the cost of his journeys. He considered that they had subscribed purely out of their Inclination of promoting Learning in General, and should there be no more Encouragers of Industry, than those who are competent Judges of the Works they promote, such Studies as are out of the common Road, would be but very slenderly provided for.

Lhwyd's old friends Anstis and Madox were so disturbed by these 'shocking expressions', which they believed might displease some and hinder others from subscribing, that they urged him to replace the entire paragraph with a note on the language of the volume and that it was 'only part of your larger design'.[29]

The new proposals of 1703 indicated that the primary purpose of *Archæologia Britannica* would now be to trace the historical origins of Britain. The historical and geographical dictionary would be the central section of *Archæologia Britannica*. Lhwyd explained the reasons for the change in the proposals and subsequently argued in the Preface to the *Glossography* that a description of British antiquities required a discussion of the ancient languages of Britain. He maintained that 'such an Essay might in this Curious Age, contribute not a little towards a Clearer Notion of the First Planters of the Three Kingdoms, and a better Understanding of our Ancient Names of Persons and Places.'[30] 'The Reason for publishing this Volume first', he airily remarked, 'will be I presume Obvious to every one.'[31] Although the change of focus might have disconcerted subscribers, Lhwyd's interests had been moving in this direction for several years. As early as September 1695 he had told John Lloyd that

upon further consideration I think it more advisable to propose the Antiquities of Wales, &c, as my main aim and designe, than the Natural History; there being so few in our Parts acquainted with this latter; and under the Umbrage of that, to collect also all the materials I can for a Natural History; which may be publishd afterwards either by my self or some other; in Latin or English. (G, p. 285)

Lhwyd and his assistants had two immediate tasks: writing the first volume 'On the Languages'; and gathering materials for the historical and geographical dictionary; as Lhwyd told Petiver in August 1704, 'You are right ... in observing that my present Business has taken me off from corresponding about Natural Things; which I shall not be able to mind; at least not in some years.'[32]

Walter Thomas, who had done so much to support Lhwyd's work, was now in very indifferent health, having suffered since 1700 from 'an oppression at my heart'.[33] Lhwyd gave him some medical advice, which Thomas appreciated as 'the greatest instances of friendshipp and affection that ever yet I met from any ... in takeing such particular and friendly notice of what I suffer in my health'.[34] Lhwyd also suggested he should consult Tancred Robinson, who refused to take a fee for his services.[35] Following the death of Thomas, some time after April 1704, Lhwyd lacked an experienced administrator to supervise the finances of the *Archæologia.*

Of the assistants who had accompanied Lhwyd on the tour, Robert Wynne resumed his studies, married and became rector of Rhiw in 1705. David Parry matriculated at Jesus College in 1701 and graduated in 1705, before beginning a career at the Museum. William Jones was immediately sent to libraries in London, where he was joined a few months later by a new assistant, Hugh Griffiths (*c.*1680–1735).[36] Their first task was to examine and transcribe historical manuscript material as sources of the planned Dictionary. Although Lhwyd directed their research by letter from Oxford, Thomas Madox warned him that 'as You are engaged in a Publique Design', his reliance on amanuenses rather than being in London himself laid him open to an 'Allegation ... to your Prejudice & to the Prejudice of your Work, that You have not seen the Records & Vouchers, Yourself', and urged him to be seen in London.[37] In London, John Anstis facili-tated the researchers' access to libraries and drew their attention to relevant documents. As on the Tour, Lhwyd trusted his assistants to judge for themselves what should be transcribed, especially when it was necessary to compare manuscript texts and printed sources. Their letters reveal the difficult working conditions for scholars at

Cotton House before it passed into public ownership in 1707.[38] Lhwyd informed a correspondent (perhaps Humphrey Foulkes) on 20 July 1701 about a major discovery:

> Will Jones is now at London a transcribing some things for me out of the *Cotton Library* & the *Tower*: in the former he has met with *Vocabularium Latino-Wallicum* written on Parchment about 200 years since wch yet is not Welsh but Cornish & so much a greater rarity but tis but brief. (G, p. 461)

In Jones's own words,

> I discovered in Vespasian A. XIV 4to Vocabular. Lat[ino] Wall[icum] on vellum that begins thus: *Deus Omnipotens Duy* ... I wonder that many of these words could tend towards Cornish since many of them are strange to me. There are three large pages and a side written as an old story [i.e. written in continuous lines] where there are names of birds and fishes and other creatures ... I cannot tell its age definitely but I believe that it could be about an hundred and fifty or two hundred years by the script. So I did not transcribe it since you have received an account of it; but I set upon the next, viz. *de situ Brecheiniauc*.[39] (translated)

Both Anstis and Lhwyd had known of the existence of the glossary, but no one had hitherto questioned the validity of the description '*Latino-Wallicum*'.[40] Though Jones may not have realised the significance of his tentative suggestion, it was not lost on Lhwyd, who told Anstis, 20 July 1701:

> [Jones] has I find made a pretty good Discovery in the Cotten Library: for one of the Vocabularies entitld Vocabularium Latino-Wallicum proves Cornish & is written on parchment about 200 years since. I have order'd him to present you with a copy of it, but not to mention it \at/ all to Dr Smith least (as is the Nature of some Librarians) he should think It too great a Rarity.[41]

Lhwyd drew upon the vocabulary extensively both in the *Glossography* and in the word lists that he drew up; in the Cornish Preface he gave his reasons for identifying the compiler as Cornish rather than Welsh or Breton.[42] He arranged for several copies to be made, for himself, for Tonkin and for John Moore.[43]

William Jones suffered considerable poverty in London. Shortly after his great discovery he complained of blisters and requested money to buy shoes and socks.[44] He returned to his lack of money in December 1701:

> Paying for washing and candles &c. &c. takes some cash from my purse every week and keeps my head below water so that I can't save a penny now (as in the summer) towards buying anything that I need. I really am very badly off for shirts and I don't have one worth wearing, and I haven't bought one since I came from Cornwall. My socks and shoes and trousers are also worn out. Travelling daily from Westminster to the Tower was a real enemy to my socks and shoes.[45] (translated)

He also wanted to buy a Welsh Bible and dictionary, and appealed to Lhwyd to authorise Walter Thomas to pay him a pound or twenty-five shillings. Relationships between Lhwyd and Jones suddenly cooled in May 1702, possibly because the latter had taken unauthorised leave. As far as Lhwyd was concerned, the breach became irreparable when Jones hurried back to Wales in June on hearing that his father was gravely ill.[46] No correspondence later than August 1702 survives,[47] and Lhwyd told Richardson on 22 December that Jones had died.[48] When Lhwyd wrote to Tonkin on 16 March 1703, he added (in a postscript tacitly omitted by Gunther), 'I have lately had a loss of poor William Jones, whom you are pleased to Remember. He died in Shrop-shire, at a Small Living the Bishop of Hereford had given him.'[49] Griffiths continued to work for a few months before leaving to complete his degree, but between February and mid-September 1705, he was again employed by Lhwyd to read the transcripts that he and Jones had made and to prepare the proper names for the Historical Dictionary.[50]

Scholars continued to visit the Museum but Lhwyd had little time to assist them. When the Swedish mineralogist and mining expert, Johan Angerstein (1672–1720), visited the Museum in September 1702 Lhwyd attended and recorded in Welsh his demonstration of assaying metal ores,[51] but was too busy to accompany him on his tour of mines. Lhwyd arranged for him to visit silver-lead mines in north Cardiganshire accompanied by his 'man', Gilli, and in October 1702 requested Tonkin to assist the 'very honest gentleman' in Cornwall (G, pp. 470–1). In recognition of Lhwyd's assistance, Angerstein presented a valuable collection of minerals to the Museum,[52] and in 1708 unexpectedly sent it more mineral specimens and 'a Stufft Rein Dear with the horns &c'. The London consignee begged Lhwyd 'to give orders for the taking away of them speedily' since 'the chest in which they are is Larg and incumbers my house'.[53]

Angerstein wrote to Lhwyd in October 1702 from Bristol en route to Cornwall, to thank him for the hospitality of Lhwyd's uncle Simon Pryse, and his aunt Elizabeth, at Ynysgreigiog, one of the Pryse houses on the river Dyfi. The visit sheds light on Lhwyd's ambivalent attitude towards his closest family. His mother's siblings, Edward, Simon and Elizabeth Pryse, seem to have been proud of Lhwyd's achievements and had always supported him. When Edward Pryse died in 1700 he left Lhwyd twenty pounds (over £3,000 in today's value).[54] Elizabeth wrote in February 1701 to say she was glad that Lhwyd had returned safely from his travels and urged him to rest; she looked forward to hearing from him as frequently as possible and to seeing him.[55] The relationship soon deteriorated; Simon wrote curtly to Lhwyd 2 October 1701, 'my sister takes it very unkindly yt you did not favour her with a letter since yr coming to England.'[56] Nevertheless, he accommodated Angerstein and Gilli for two nights at Ynysgreigiog in October 1702 and took them to visit the Esgair Hir mine. When Simon sent Lhwyd an account of their visit, he added that he and Elizabeth had each entrusted Gilli with a crown (five shillings, worth well over £40 in today's value) to give him, and expressed his disappointment that Lhwyd had not been able to make the journey himself. Elizabeth wrote at the beginning of December hoping to hear from Lhwyd, who had not acknowledged the gifts.[57] After a long silence, by mid-August 1703 her patience was finally exhausted:

Cousin Lloyd

I wonder and admire that you doe soe little value us to shew such slights to your nearest relations, to your owne Uncle your mother's brother, and your Aunts your Mother's Sisters, which I find by you that you little regard your Mother's relations. This is the fourth Letter which I write unto you since the gentleman and your man came into wayels into Cardiganshire which is now above 3 quarters of a yeare ... and wee received them to our howss and mayd them welcome upon your account what the countrey could afford and Lodgd them with us to or 3 nights and my brother and my self sent each of us a small token being Crowns a peece unto you with 2 Letters with them, and you had not the civility nor respects to write unto us whether you received them or not ... if you have a mind to strang your self and keep off from writing unto us you may doe what you please I have nothing to add but my well wishes and respects unto you.[58]

When Elizabeth died in 1706 she bequeathed ten pounds (over £1,600 in today's value) to 'Edward Lloyd Gentleman, the sonn of my sister Bridgett Pryse deceased'.[59] These letters are in marked contrast to the correspondence between Lhwyd and other family members with broader horizons – John Parry between 1698 and 1704, and his kinswoman, Sage Lloyd of Carmarthen, in 1703.[60] Parry and Sage Lloyd were intellectually minded people with wide interests; indeed Sage Lloyd may have been the only woman (apart from the Duchess of Beaufort) considered by Lhwyd to be a rational being. Lhwyd was always focused on his work, and sometimes appeared ready to exploit relationships to his own advantage and drop those who had no more to offer.

Lhwyd visited Cambridge for three or four weeks in the summer of 1702, lured there, he told Henry Rowlands, by a 'false title ... in their late printed catalogue[61] wch promis'd me a map of Britain and Ireland, by Giraldus Cambrensis' (G, p. 472). He stayed with an old acquaintance, William Vernon (1667–1715), a Fellow of Peterhouse and enthusiastic entomologist and fossil-hunter, and 'drained [him] of fossils and insects'.[62] At Cambridge Lhwyd discovered that the Juvencus manuscript, Cambridge University Library, MS Ff.4.42, contained Old Welsh glosses and a series of Old Welsh *englynion*. Nine of them, a single poem, were written on the margins of folio 1 and three others were written as a single line on the top margin of folios 25–26.[63] For some reason (perhaps because he had not noticed them) Lhwyd paid no attention to the nine *englynion* which, as Christian poetry, he might have found less difficult to interpret.[64] As Hearne recorded in 1730, Lhwyd succumbed to temptation and cut out the three *englynion*:

> Mr [Thomas] Baker had the same opinion of Mr Edw. Lhuyd that I have (as a plain, open, hearty man, *sine fuoco* [without fire], till he came to Cambridge where, being trusted in the public [University] Library, he met with a Juvencus MS … & there being some antient British notes … in the Margin, he cut 'em out with a penknife.[65]

They remained in Lhwyd's possession until his death but were then quietly returned to Cambridge by Humfrey Wanley.[66] Lhwyd was apparently driven by a desire to examine the *englynion* at his leisure, since he could not yet interpret them: 'nad ydui etto yn y dealh'.[67] Contemporaries to whom he sent copies were baffled; William Baxter (1650–1723),[68] in Lhwyd's opinion 'a person of learning and integrity, thô, I fear me, too apt to indulge fancy' (G, p. 176), concluded that the last two stanzas were 'a plain praediction that our gracious Queen will yet have a Prince of Wales'.[69] When sending Henry Rowlands a transcript of the verses, Lhwyd remarked that 'a famous linguist and critick' had supplied 'such an interpretation, as I shall not trouble you withal' (G, p. 473), adding drily in a subsequent letter that the reading 'will, I believe, surprise you, as well as it did me, when you will see it' (G, p. 473). Although the verses were clearly 'British', it was not clear that they were Welsh, Lhwyd inclining towards 'Strathclwyd Welsh or Pictish'. He was, however, able to make some comments[70] and expressed his initial thoughts on the orthography of the glosses and the verses to Humphrey Foulkes in December 1702 (G, pp. 474–8). Lhwyd's visit to Cambridge marks what has been called 'the discovery of Old Welsh',[71] though it was not recognised as such at the time. Although Lhwyd had previously seen 'old' (medieval) manuscripts containing early Welsh poems, he had not hitherto encountered any 'very Ancient Manuscripts' (pre-twelfth century) comparable to those studied by contemporary scholars of Old English such as Hickes. Lhwyd realised that such manuscripts, if they could be found, would serve as a link between inscriptions and later manuscripts. Shortly after the Juvencus discovery, in September 1702, Humfrey Wanley brought to Lhwyd's notice the Old Welsh marginalia in the Book of St Chad or Lichfield Gospels, a manuscript that had recently been 'maimed by the ignorant Bookbinder' who had 'cut away a great part of those Marginal Entries'.[72] As he told Lhwyd in December, he believed these to be 'the most Antient Remains of the British Language in Writing, that are now to be found in England'.[73] Wanley also drew Lhwyd's attention to two Latin manuscripts in the Bodleian (Ox. 1 and Ox. 2) containing Old Welsh material which Lhwyd had seen, but whose significance he had not appreciated. His correspondence during these years reveals much of his thinking about linguistics; he now paid more attention, especially in his letters to Henry Rowlands and Humphrey Foulkes, to phonological and lexical correspondences between languages, in comments that foreshadow the fuller discussion in the *Glossography*. It was also during this time that Lhwyd began to buy more Welsh books. Some were found for him in London by Thomas Dafydd, 'a Cardiganshire Day labourer about London' (G, p. 505), whom he had first met in 1699.[74] Dafydd's brother Griffith Davies undertook errands for Lhwyd in London, and his cousin Hugh Griffiths was one of Lhwyd's amanuenses.

In the autumn of 1702 Lhwyd withdrew from Oxford to concentrate on his book. In mid-November he told Lister: 'I was about six miles hence (where I now spend much of my time because of some unnecessary interruptions at the Museum) when your present came' (G, p. 473); his retreat was the hamlet of Appleton to the south-west of Oxford (G, p. 478). Friends and colleagues understood his need for undisturbed study, as Jacob Bobart explained to Richardson in August 1704,

Mr Lhwyd frequently retires (for the convenience of private studies) some miles off. His work goes on, tho' not with that celeritie
as could be wished; some of his friends being sorry that he is so
deeply engaged in antiquities and the nice prosecution of language
matters, which prove very knotty. (G, p. 449)

The University Printer began work on the book 29 September 1703
'for the author' – that is, Lhwyd was to bear the cost of printing and
publishing (G, pp. 527–8). The press faced a considerable task as the
Glossography comprises four hundred and sixty-four pages, virtually all
of which contain three columns of text in several languages and employ
a wide range of type-faces and symbols. Progress was slow since, as
Lhwyd explained to Thomas Smith in March 1704,

My compositor is now on the eleventh sheet of my Book. He does
but one sheet a week but 'tis a small letter and contains almost as
much as two of such Print as Dr. Plots' *Histories* &c.
  I would put two upon't but they have not Letters enough of
the sort; especially not Capitals no Dictionary having been ever
printed at the Theater excepting the Saxon, which as you know was
done without Capitols & in continued Lines. (G, p. 500)

Printing was also slow because the press was producing another major
work, *Thesaurus Linguarum Septentrionalium* by George Hickes (1642–
1715). Hickes told Lhwyd that he was surprised 'that [Edmund] Bush,
who was the printer of my book, should become the printer of yours,
the effect whereof ... would be the retarding of both'.[75] In August
Lhwyd complained to Thomas Smith (1638–1710) that 'my Printer
has been necessitated to leave mine, as often as he should receive more
copy of that: However there are now 22 sheets finishd of my Irish–
English Dictionary which ... makes about a fourth Part of the book'
(G, p. 501). By mid-December he told John Lloyd that although he
had provided sufficient copy 'there is not \much/ above a third part of
it as yet printed ... the Delay is wholy owing to the Printers, who will
always have several irons in the Fire & also keep holyday when they

please' (G, p. 503). The first part to be set was the final section of the book, the Irish–English Dictionary. This allowed Lhwyd to send proof-sheets to several scholars; the additions and corrections of the only two who responded, one from Scotland and Roderick O'Flaherty from Ireland, appeared in an Appendix.[76] The piecemeal publication process enabled Lhwyd to expand certain sections. As he explained to Bishop Humphrey Humphreys mid-October 1706, 'The Effect of this Tedious printing, has been that having so much spare time on my Hands, I made the Irish Dictionary and some other Tracts whilst they were a printing, twice as large as I design'd.'[77] One consequence of this expansion was that the Cornish dictionary, now NLW, Llanstephan MS 84, became too extensive to be included in the first volume.

The *Glossography* was published in May 1707. In June Lhwyd told Richardson that the names of a few subscribers, including his own, had been accidentally omitted but were subsequently 'printed & pasted in about half the Impression', which Lhwyd thought would serve until the full list appeared in the projected second volume.[78] Humphrey Foulkes had warned Lhwyd in April 1706 that 'your 2nd vol of Customs would have been much more acceptable to the Gentry, to be sure you'll prom-ise that \pleasant & entertaining vol./ shall follow very suddenly this dry and jejûne one',[79] but Lhwyd told John Lloyd in August 1707 that 'The Linguists & Antiquaries in these Parts are so well satisfied with this Volume that it sels much beyond what could be Expected of a Book so Forreign' (G, p. 532). Baxter published an enthusiastic account of the book in the *Philosophical Transactions* for September 1707.[80] Lhwyd had 'a few supernumary sheets' of this printed and added them to the unsold copies of the book.[81] Other copies of Baxter's notice were sent to friends; Henry Rowlands asked for '6 or 7 of them to be inserted into the Bookes I pass'd off to my friends'.[82] When sending copies to John Lloyd, Lhwyd observed that if he passed them on to subscribers 'perhaps they'll entertain a more favourable Opinion, than they have hitherto.'[83] One of the few to appreciate fully the scope and nature of Lhwyd's achievement was George Hickes, who wrote on 2 June:

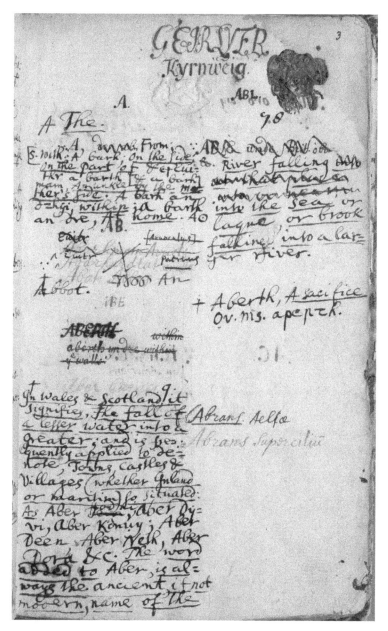

FIGURE 8 Lhwyd's Cornish Dictionary, *Geirlyer Kyrnwëig*, NLW,
MS Llanstephan 84, intended for inclusion in the *Archæologia*

The moment I received it, I sat down to peruse it for 4 houres together, and had sit longer at it, if other affaires had not called me of ... it will be very satisfactory to all men, who have a genius for antiquity, and the more Learned and iudicious they are, the more they will approve it ... I wish for my own sake ... it had been printed 20 years ago, so usefull would it have been unto me. It hath cost you a great deal of thought, and by consequence required a great deal of time, and now I consider the Harmony, and accuracy of the severall parts of it, I wonder we had it so soon.[84]

Hickes had understood Lhwyd's aims better than did Gunther over two centuries later, when the latter described the *Glossography* as 'a curiously composite production, by several authors in several languages' (G, p. 451). The book is very much Lhwyd's work but some important sections were entrusted to his 'pupils', the 'Armoric Grammar and Vocabulary' (Chapters 3 and 4) being a translation by Moses Williams (1685–1742) of the book by Julien Maunoir acquired 'too expensively' in 1701. The 'Essay towards a British Etymologicon' (Chapter 8) was written by David Parry, probably from material supplied by Lhwyd. Lhwyd's ultimate objective was to compare ancient British personal and place-names with linguistic evidence from Gaul, but for such a comparison to shed light on the relationships, origins and migrations of peoples, it was imperative that the grounds of the comparisons should be properly established. Lhwyd laid great stress on the ability to distinguish between native words, which might reveal the affinities of languages, and loan words, which could indicate language contacts and the movements of the speakers of these languages. Historians unfamiliar with the ancient languages of Britain should use linguistic evidence in an informed manner, based on principles of phonetic analysis and correspondence, rather than relying on chance similarities in sound between words plucked from dictionaries. For Lhwyd, linguistics, and etymology in particular, was not a study to be undertaken for its own sake; it was, rather, part of the historian's equipment. The first volume of the *Archæologia Britannica* project, therefore, needed to include glossaries and grammars, phonetic correspondences, and also, in Chapter 7, a catalogue of Welsh and

Latin 'British' manuscripts, which listed the main literary and linguistic sources. This first volume is a tool-kit for users of *Archæologia Britannica*.

The *Glossography* was an attempt to 'methodise', to impose analytic order upon the correspondence of languages one with another.

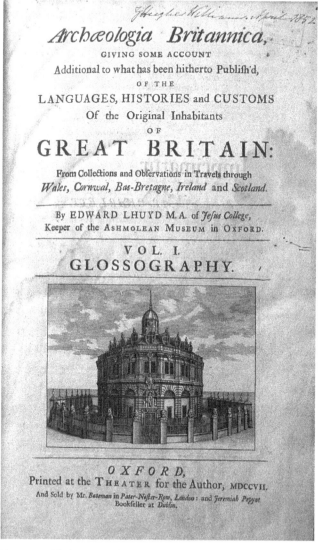

FIGURE 9 Title page of *Archæologia Britannica* (Oxford, 1707)

It contains two types of linguistic material: formal grammars of Irish, Breton and Cornish with glossaries;[85] and discussions and examples of etymologies. Lhwyd had realised from the first that his 'History of Wales' would have a strong linguistic base. As early as 1688, in the comparative glossary of Irish and Welsh words that agree in 'sense and sound' discussed in Chapter 4, he was beginning to adopt a systematic approach to questions of language. He was concerned with the broad question of the origins and interrelationships of languages; by examining individual words he attempted to deduce evidence which could be used, together with written sources, for tracing the movements of peoples. Etymologising was a common tool in such historico-linguistic studies, and Lhwyd could draw on a great deal of sixteenth- and seventeenth-century writing on these matters by scholars like Vossius (Gerrit Janzoon Vos, 1577–1649), Gilles Ménage (1613–92), Stephen Skinner (1623–67), Pierre Besnier (1648–1705), Edward Bernard and Hickes. Lhwyd's emphasis on borrowings and cognates as explanations of similarities between languages and his attempts to devise a rational methodology to distinguish between them place him in the mainstream of seventeenth-century developments and make him a significant contributor to contemporary linguistic thought. His early methodology had served to state the relationships of single words within an accepted framework, based on historical evidence, of related languages; the need now was to establish these relationships on linguistic grounds. By the time Lhwyd came to write the *Glossography* he had turned his attention to a detailed consideration of linguistic correspondences, most especially of the rules governing phonetic change. Two of the ten sections of the book are devoted to etymology (a third, Section II, is *A Comparative Vocabulary* of the Celtic languages which also notes cognate forms), and both of these, I, *Comparative Etymology or Remarks on the alteration of languages*, and VIII, *An Essay towards a British Etymologicon*, contain discussions of the criteria which Lhwyd employed to determine correspondences between compared words. Real or assumed similarity of sound and sense had led to an unregulated 'correspondence of letters' by which scholars altered orthographical forms according to their theories of word composition rather than through a methodical analysis of phonetic correspondences.[86] He aimed to reduce the arbitrary element in etymology by formulating a set of what we would call rules of phonetic change which were based on phonetic groups. Lhwyd was concerned both with *how* words in different languages should be compared, and also with *which* words may properly be examined. His guidelines for etymology are well formulated: an identity or analogy of signification is always required; declensions and conjugations are proper to individual languages; old or obsolete words may be found not only in ancient texts but in living modern dialects; the words to be compared should be the oldest and most simple, or the obvious and necessary. By 1700 Lhwyd's views were in the mainstream but his specific guidelines for identifying phonetic correspondences were innovative and mark him as a linguist more interested in evidence than in hypotheses.

Lhwyd's six principles of the 'Origin of Dialects' which became in time 'distinct Languages', can all be found in the works of his immediate predecessors, as he himself implies.[87] His great analytical gifts, perhaps his experience as a botanist and zoologist concerned with the typological classification of species and genera, allied to a phonetician's ear, led to his working out in detail of a host of examples of these six principles under twenty-four observations, which list correspondences between the Celtic languages, Latin, French, Spanish, Italian, English (and English dialects) and sometimes German. These are then summarised under ten headings, including ten classes of phonetic change. The most interesting of these is 'Permutation or Change of Letters'. These 'permutations' are truly phonetic. They are brought about 'by the ear of the multitude not from writing', and they are gradual ('by degrees') over a period of time. One of his examples is the change from K in Celtic and Latin to H in Teutonic languages, and this is analogous, he says, to the change from northern Welsh initial C pronounced 'very gutturally' (that is, *chw-*) to southern Welsh H, as in *chware – hware*. The pattern of initial mutations (mutations of initial consonants) in the Celtic languages suggested to him an explanation of phonetic change in the development of other European languages. These phonetic changes he regarded as the best guidelines for tracing language relationships, and he lists those which he regarded as being the most significant. Lhwyd, more than any other etymologist, attempted a rigorously reasoned analysis of phonetic change and correspondence, and these principles are a major contribution to the comparative method.

The most significant of these for Celticists is the correspondence which Lhwyd noted in 1703 between British P and Irish C (or Q):

> for you must know I am troubled with an Hypothesis of C. Britons and P. Britons, the C. BB would begin no word with a P, and therefore for pen a head they said Cean or Kean ... In so much that I cannot find the Irish have a word (purely their own) that begins with a P. and yet have almost all ours, which they constantly begin with a C. (G, p. 491)

Lhwyd's aim in the *Glossography* was to trace the 'Original Language' of the British Isles. The affinity of Welsh, Cornish and Breton was in no doubt, though the phonetic bases of the relationship had not been established before Lhwyd's work, nor had the degree of similarity between them been noted:

> I ... found that the Armorican and Cornish differ'd less than the present English of the vulgar in the North from those of the West of England; but in respect of us the difference is greater ... Their language [of the Cornish plays] comes nearest that dialect of the British, call'd in Dr. Davies Gwen Nwyseg, or the Language of Monmouth and Glamorgan. (G, p. 441)

Although earlier scholars, notably George Buchanan (1506–82), Boxhorn and Camden, had argued for the affinity of British and Gaulish (or what was called in the seventeenth century 'Celtic'), Irish had not been fully integrated into this scheme until Lhwyd demonstrated how Gaulish words, place-names and personal names might be reflected in Irish. However, when he began to compose a list of Irish words which agreed with the British languages, he also discovered that Irish bore a closer affinity with them than with Gaulish, a view which his etymological rules confirmed. Irish and Welsh, he concluded, were the first and original languages of the British Isles. Both were linked to Gaulish and had a common origin with Latin and Greek.

The 1695 *Design* had shown that the question of British origins lay at the root of Lhwyd's linguistic work. His research and methodical analysis over the next ten years led him to a view of the interrelationship of the Celtic languages. In the Welsh Preface to the *Archæologia* and in his correspondence with a few friends he set out a theory of Celtic migrations to Britain, which he had been developing since about 1700, though he was not prepared to discuss the more distant theme of the ultimate mother language.[88] The C (Irish) Britons were, he believed, a Gaulish colony in Britain, whose priority in time is shown by the existence of place-names containing the element *isca, ex*, cognate, so he claimed, with Irish *uisge* but not found in British; a second wave of Gaulish settlers, P Britons, forced them northwards to the Highlands and the north of England, where they later became mixed with Scots and English, and to Ireland. The place-names of the Lowlands, he noted, were more akin to Welsh than those of the Highlands. The C Britons comprised a distinct nation in Ireland, Gwyddelians, whose language was akin to Welsh. However, another nation in Ireland was the Scots (some of whom migrated to the Scottish Highlands), in Welsh *Skuidied*, in Irish *Kinskuit*, who, Irish tradition[89] and Nennius had said, came out of Spain. Lhwyd claimed to have found an affinity between part of the Irish lexicon (but not the British) with 'Old Spanish' ('Cantabrian' or Basque) which confirmed this tradition.[90] In the Welsh Preface to the *Glossography* he noted a few Basque words together with the Irish, and enlarged upon the relationship of Irish and Basque. He ascribed a Germanic element in Irish to Belgic influence, reflected in Irish stories about the Fir Bolg. Many strands in this theory failed to gain currency since it remained hidden in the Welsh Preface. Lhwyd was, however, proud of the 'novelties' of his argument and, while acknowledging that he had no written authority for his views, was prepared to justify them on strictly linguistic grounds: 'All the arguments us'd is the agreement of languages.'[91]

Lhwyd had long been interested in Pezron's ideas.[92] At the beginning of September 1703 he told Lister he had seen a copy of Pezron's book, *L'antiquité de la nation, et de la langue des Celtes, autrement appellez Gaulois* (Paris, 1703),[93] and at the end of the month sent John Lloyd a sceptical summary of Pezron's views, preceded by the mordant comment that Pezron had 'Infinitely outdone all our Countreymen as to National Zeal'. At the end of his book, Pezron appended three lists of words: 'those the Æolians & other Greeks borrow'd from our Ancestors the Titans ... those the Romans borrowd from the Umbri ... those the Germans had from ye several Gaulish Colonies planted in their Countrey' (G, p. 489). Chapter 2 of Lhwyd's *Glossography*, 'A Comparative Vocabulary of the Original Languages of Britain and Ireland', might, at first glance, appear to resemble Pezron's lists of words similar in appearance and meaning, but Lhwyd's list followed his carefully argued Chapter 1, 'Comparative Etymology', which was intended to show the affinity of British with other languages. Where Lhwyd was setting out the grounds for comparisons, Pezron could see only borrowings from Celtic, mainly from modern Breton.[94] Lhwyd's search was for the most ancient languages of Britain, which he was content to call Celtic, and for their speakers, the original settlers, but Pezron attempted to integrate his British languages into the biblical scheme of Genesis 10 and ideas about the origins of nations. Pezron considered Celtic to be the ancestor tongue from which the major European languages were derived, and claimed that modern Breton and Welsh preserved this ancient language in its purest form:

> they & we are the onely Nations in the world that have the Honour to have preserv'd the Language of Jupiter & Sadurn ... Princes of the Titans the progenitors of the Gauls, & to have had an Empire from the Euphrates to Cape Finister in the time of Abraham. (G, p. 489)

Lhwyd could agree with Pezron's views on the antiquity of the Celts in Britain, but he could not follow his speculative leap from Gauls to Gomerians, Cimbri, Titans and other predecessors. The scholarship and suspicion of theorising that had earlier led Lhwyd to reject Woodward's flights of cosmology now prevented him from accepting Pezron's unsubstantiated lexical prehistory.[95]

The achievement of the *Glossography* is twofold. Lhwyd provided etymology with a rational basis within the conceptual framework of seventeenth-century scientific thought, thereby setting the comparative method on firmer ground. One consequence was that he justified the use of the term 'the Celtic languages' as a meaningful one and through this gave credence to the concept of a common Celtic culture.[96] Lhwyd realised that there was a common linguistic context to early Britain; all its languages were related and this linguistic commonality reflected a common cultural context. Lhwyd did not suffer from 'Druidomania', but he did believe that the druids represented a common religion to which the archaeological finds – stone circles, cromlechs and the like – might testify, just as the so-called snake stones might reflect a common superstition. As shown in Chapter 9, at least one verse form found in both Welsh and Cornish was, he believed, common to all Celtic cultures and might have been used by druidic teachers. When Lhwyd began in 1698 to refer to Celtic languages on linguistic grounds, though perhaps with implied ethnic implications, he was laying the foundations of modern Celtic studies.[97] It must be noted that his use of 'Celt' was adjectival; he wrote of Celtic languages and of a Celtic alphabet, but apparently never wrote specifically about 'the Celts'.

Once the *Glossography* had appeared Lhwyd could turn to the daunting task of 'methodising' the materials for his next volume, the Historical Dictionary. William Jones and Hugh Griffiths had been transcribing sources in London from 1701 onwards. Lhwyd, assisted by David Parry and Moses Williams, would have undertaken the final editing of the Dictionary and presumably he would have written an introductory essay. Had he lived a little longer, the Historical Dictionary would have been ready for publication, as perhaps would the entire second volume of *Archæologia Britannica*. Griffiths grieved that he could not assist him in the work, telling Lhwyd in June 1707:

> I cannot without great regret ... think of ye infinite trouble and pains you must be at to publish ye 2nd volume; I could heartily wish my self in such a place where I should not have such occasion

for weekly Composing as here I am obliged to. I should then very willingly be at your service, either to transcribe ye Collections out of Dr Powel, or any other thing wch you might think me capable of serving you in.[98]

Possibly as a relief from the crushing task, Lhwyd told Richard Mostyn in November 1707 that he was considering adding an 'Onomologia Britannica \(or Celto-Britannica)/ or a dissertation on the method of nameing persons and places, us'd amongst the Gauls and Britans' to the edition of *Cæsar's Commentaries* which 'Dr. Aldridge' proposed publishing.[99] He believed that 'There has been nothing of that kind, that I know of as yet attempted: and perhaps the learned part of my subscribers would think that part of my Time as well bestowed as any' (G, p. 535). The idea was still alive in October 1708 when Lhwyd told Richardson he intended to publish 'an 8vo \in Latin/' on the subject and asked him for Yorkshire mountain and river names 'not to be found in the Maps',[100] but nothing was heard of it subsequently.

The last work that Lhwyd saw through the press was an extended review, published in the *Philosophical Transactions* for 1708, of J. J. Scheuchzer's *Itinera alpina tria*.[101] Lhwyd had agreed to prepare this in April 1708,[102] but its publication was delayed, since the Oxford wagon caught fire en route to London in early August and all its cargo, including Lhwyd's paper, was destroyed.[103] Since Lhwyd had not kept a copy, John Thorpe (1682–1750), editor of the *Transactions*, told him that Robinson had suggested he should dictate his text to save time; Thorpe had left room for it in the next number.[104] Lhwyd managed to reconstruct his review quickly enough for Robinson to write on 25 September that it was 'now printing'.[105] Most of the review was an objective account of the book which was generally considered to be more useful than the original, since Lhwyd had provided a systematically organised account of a rather diffuse work. He did, however, seize an opportunity to attack once again Woodward's theory of an all-dissolving Deluge when discussing the Sennenstein, in fact the Säsagit,[106] an impressive natural stone pillar near Kunkels. Lhwyd commented:

[Scheuchzer] says, it is not (as those of Stonehenge and divers other Places throughout Britain and Ireland) erected by Mens Hands, but natural ... This one would think scarce reconcilable with an opinion he is said to maintain of an Atomical Dissolution of all Things the Terrestrial Globe consisted of at the Deluge; for if so, we are left to seek what Mould such a Pillar should be cast in ... But the truth is, he has no where hitherto, that I know of, profess'd publickly that opinion, which has been long since sufficiently exploded in the Ingenious Examination of it. For in his Epistle before the Translation of Dr. Woodward's Essay, he only tells him, that his Book had convinced him, the Fossil Shells, &c. were of Marine Origin; which amounts to no more than what I had publish'd in the *Philosophical Transactions* two Years before that Essay appear'd.[107]

Woodward rose to the bait, as Thorpe gleefully told Lhwyd towards the end of November 1708:

Some days after the publishing the last Transaction, Dr W–d sends several Messages to me to wait on him ... to know Why, and wherefore, and for what Reasons, and by whose Order, I publish'd such a scandalous Reflection on him in the last *Transaction*. But I not returning a satisfactory Answer, he goes to the President with heavy Complaints against the Secretary and me ... yesterday it was debated in Council: where it was agreed that the *Transactions* were not the Act of the Society's, but the Secretary's, who might put in what Papers he pleased ... Afterwards Dr W– (not knowing what had passed in Councill) begins his Tale at the Meeting with all the Eloquence and aggravating Circumstances imaginable. The Secretary ... told him at last, that not he, but the Authors only were accountable ... but, not satisfied with this Answer, he talked of nothing but having Justice, and Satisfaction, and taking Courses ... We are yet uncertain how this Controversy (which gave us a great deal of Diversion) will be decided, whether by the Sword or Pen: If the former, Mr Mead has promised to be

Dr Sl[oan]e's second, and Mr Oliver mine: I would have you likewise stand on Your Guard, and timely provide Yourself ... Not withstanding all this Scuffle, I propos'd You to be a Member ... on Tuesday Next [you] will be elected ... to the Great Mortification of Dr W–d.[108]

Woodward, enraged by the Scheuchzer review, attempted to prevent Lhwyd from being elected a Fellow of the Royal Society.[109] Thorpe told Lhwyd on 30 November 1708:

This morning at Gresham College You was approved in Councill; and being after proposed in the General Meeting, Dr W took occasion to make a very long Speech against Your Election, but being answer'd very smartly by several Members, and the Dr finding no body to second him ... You was chosen with a very inconsiderable opposition.[110]

Even after this humiliating defeat Woodward's hostility continued unabated. Nicolson remarked in his diary, 4 April 1709, that Woodward was, 'Hard on Mr Lhwyd, Dr Richardson and Mr Rowlands', all critics of his theory,[111] and Hearne noted a month later that Woodward complained that Lhwyd had 'given so many Samples of a Malice [amounting?] to hatred' against him.[112]

In January 1709 Lhwyd was 'encouraged' to stand for election as Superior Divinity Beadle. The post, a lifetime appointment with an annual stipend of about a hundred pounds, was no sinecure since it involved considerable administrative duties. Hearne withdrew his own nomination in favour of Lhwyd, who was elected 11 March 1709 with a majority of twenty votes over his only rival, John Colinge[113] of New College. Although Hearne had secured Lhwyd the votes of several nonjurors, this did not necessarily indicate Lhwyd's political sympathies; Hearne had made it clear that he stood aside because Lhwyd was 'a man of far better merits than I can pretend to, and withall because he is my intimate Friend'.[114] The costs of campaigning aggravated Lhwyd's immediate financial difficulties but had he lived longer his stipend would have finally resolved them.

Although he drove himself hard, Lhwyd was a sociable man. He was a (or maybe 'the') founder of the Red Herring club in Oxford in 1694.[115] A pen picture of Lhwyd, written around 1756 by a 'rev. Mr Jones, a curious collector of anecdotes', says that

> At evenings, after his hard study in the day-time, he used to refresh himself among men of learning and inquiry, and more particularly Cambro-Britons, in friendly conversations upon subjects of British antiquity; communicating his extensive knowledge therein, with much good humour, freedom and cheerfulness, and at the same time, receiving from them farther and more particular informations, subservient to his great and laudable designs.[116]

The young Welshmen in Lhwyd's circle may have made him more aware of popular Welsh culture during his last years. About 1707–8 he composed a set of *englynion* commemorating a standing jump 'from the museum steps to the adjacent pillar', a distance of some three and a half yards, made by Huw Puw (d. 1743), who had been one of Lhwyd's fellow-undergraduates at Jesus. Lhwyd decided to preserve the memory of the feat in verse for a new generation of students. The *englynion* were carved on a pillar, 'one of the Buttresses in the court of the Ashmolean Museum' and, according to Hugh Davies (1739–1821), who was then at Jesus, remained legible in 1760. A recent search failed to find any trace of the verses or even identify the pillar.[117] A further example of Lhwyd's involvement in the culture of this circle of Oxford Welshmen was his role in the publication of *Hoglandiae Descriptio*, a reply to *Muscipula* by Edward Holdsworth (1688–1747),[118] a mock-heroic Latin poem deriding the proverbial fondness of the Welsh for cheese.[119] Thomas Richards (1687?–1760), who had matriculated at Jesus College in 1708, replied in his *Hoglandiae Descriptio*, a mock-epic based on the idea that Holdsworth had been born in Hampshire, a county famous for its pigs. It has been claimed that Lhwyd suggested the theme of *Hoglandiae* to Richards and composed a Latin preface which was considered 'too severe' and replaced by a milder version. An English translation of the original Latin preface was used for the

English version, *Hog-Land*, published in 1711. Concrete evidence of Lhwyd's part in the affair is that a sketch for the woodcut of a gigantic boar printed on the title page is amongst his papers.

Lhwyd claimed to have enjoyed good health for much of his life. When briefly ill in November 1692 he told Lister: 'I have not been accustom'd to take purges or any other physick; this being but ye third time, that I can remember ever to have taken any' (G, p. 170), and in May 1698 told John Lloyd (who had heard a false report of Lhwyd's death) that he had 'not been one day very sick these ten years; nor have I ever enjoy'd my health (God be thanked) better than in my Travails' (G, p. 372). Within three years, Lhwyd's health had deteriorated; as Lloyd wrote in May 1701, 'We were in no small fear here, lest we shd loose you. For considering yr old Cough & Bala headach, ye different airs & climates you travail'd through, ye multiplicity of business & studies you were continually overwhelmed with ... we gave you for gone.'[120] It is not known when Lhwyd's asthma first manifested itself, but he began to suffer recurring, incapacitating headaches during the Welsh tour. In November 1700 Robert Wynne hoped that he was 'rid of his severe headake' and was fit to sail for Brittany.[121] The headaches continued after Lhwyd had returned to Oxford, Lister telling him in September 1703, 'I hope you are better of the head ake.'[122] Robinson believed that the headache was caused by Lhwyd leaning his 'head too low & too long to write if not to read'.[123] During his last few months Lhwyd seems to have spent much time working in his study in the damp ground floor of the Museum, and it was there that he died, aged forty–nine, 'Wednesday ... between 10 and 11 Clock in the Evening', 29 June 1709, according to Hearne's first account,[124] the cause of death being pleurisy, brought on by the asthma that Lhwyd had suffered for many years.[125] Hearne later put the death on Thursday 30 June, 'about one o'clock morning'.[126] Lhwyd was buried the following day in the parish church of St Michael 'in the South or Welsh Isle, as it is called, being the burial-place of the members of Jesus-College ... attended by the members of the common room of Jesus-College, and the Beadles'.[127]

There was no marker on Lhwyd's grave. In 1713 John Thorpe spoke of setting up a monument to him, telling Hearne that 'some Persons here in London' seemed very willing to contribute towards the cost. He believed that James Harcourt (1680–1739), a fellow of Jesus, had 'a Draught of his Face &c'. Hearne replied that Harcourt did possess a picture 'of my late excellently learned & equally modest Friend', but it was 'taken after his Death, and … not at all like him'.[128] No more was heard of the project, nor of Harcourt's portrait of Lhwyd. The only near-contemporary portrait known to survive is a pen and ink drawing in the Ashmolean Museum Book of Benefactors, forming the background of the illustrated capital 'E' above the account of 'EDVARDUS LHUYD'. Since David Parry wrote the account of Lhwyd, the posthumous image, by an unknown artist, was made sometime between 1709 and Parry's death in 1714. This became the model for a series of later portrayals of Lhwyd, in which the original likeness of a real person, fond of the good life perhaps, self-confidently looking directly at the viewer but with the hint of a smile and some uncertainty, became increasingly stereotyped and lifeless with each iteration.[129]

11

AFTERLIFE[1]

A s early as 1695 Lhwyd had been concerned to ensure his work would be completed should he die prematurely (G, p. 293). His choice of successor was singularly unfortunate; virtually any of his amanuenses would have been more suitable than David Parry. Although Parry was 'little inferior to his predecessor Lloyd in natural history or in the knowledge of Cambrian, Anglo-Saxon and other languages', according to Uffenbach, he was 'always lounging about in the inns' and could not 'show strangers over the museum for guzzling and toping'.[2] Parry's weakness was known before Lhwyd's death, but Lhwyd was unaccountably prepared to tolerate fallible people such as Parry and Gilli, while dismissing others, such as William Jones, for apparently trivial misdemeanours. Parry's death in December 1714 terminated any prospect of continuing Lhwyd's researches at the Museum.

Immediately following Lhwyd's death there was considerable speculation whether the second volume of the *Archæologia* existed in publishable form. Thomas Smith asked Hearne on 8 July 1709 whether Lhwyd had finished the volume and what 'curious papers' he might have left.[3] Hearne replied a few days later that Lhwyd 'had not digested the materials for Vol. II', a view which seems to be at odds with suggestions that Lhwyd had produced a draft of the geographical and historical dictionary.[4] Replying to Hearne, Smith expressed his regret but hoped that Dr Hudson (the Librarian) would secure Lhwyd's papers for the Bodleian.[5] This was not to be, as Lhwyd died in debt and intestate.[6] Since his printed books were sequestered by the University to pay the outstanding costs of printing the *Glossography*, many were preserved in the Bodleian Library.[7] Parry does not appear to have attempted to preserve Lhwyd's papers, and most of them were scattered. Some correspondence was retained at the Ashmolean, perhaps because it was regarded as part of the Museum archives rather than Lhwyd's personal effects, and other letters subsequently made their way there, many of them through the efforts of William Huddesford (1732–72), Keeper from 1755 onwards.[8] The natural home for Lhwyd's other papers would have been Jesus College or the Bodleian, but neither attempted to acquire them. Robert Harley (1661–1724), whose librarian Humfrey Wanley was well aware of their value, hoped to buy them.[9] To Wanley's disgust, the collection was bought in 1715 by a former Jesus man, Sir Thomas Sebright (1692–1736), fourth baronet, of Beechwood, Hertfordshire, an enthusiastic book collector. As was shown in Chapter 9, Lhwyd's Irish manuscripts were donated in 1782 by Sir John Sebright to Trinity College, Dublin. Another large (Welsh) part of the collection was given by Sir John Sebright (1767–1846), seventh baronet, to his cousin by marriage, Thomas Johnes (1748–1816) of Hafod, Ceredigion, in 1796–7, but most were lost in the Hafod fire of March 1807. Descriptions of the remaining Sebright manuscripts in the April 1807 sale catalogue, indicate the range of Lhwyd's work. Among the purchasers at the sale was Sir Watkin Williams Wynn III (1767–1840) of Wynnstay, his acquisitions including 'Thirty-six of Edward Lhwyd's Pocket Memorandum Books of Observations on Natural History and Antiquities' and thirteen volumes of 'Notes and Drawings of Antiquities, Monuments, &c in Wales'. Most of these were lost when John Mackinlay's bindery in Bow Street was destroyed in the Covent Garden Theatre fire 30 September 1808, and the survivors perished in the Wynnstay fire of March 1858.

Some Lhwyd material was published following his death. Those additional notes not included in the 1695 *Britannia* were printed in the 1722 edition.[10] Sloane sought unpublished letters for *The Philosophical Transactions* from Richardson and others (G, pp. 553–4). In his *Glossarium Antiquitatum Britannicarum* (1719), Baxter published Lhwyd's notes on British place-name elements, preceded by his letter to Nicolson,[11] and added two pages of scraps from Lhwyd's unpublished letters to the second edition of 1733. A few remarks by Lhwyd on Cornish inscriptions and monuments, mainly from two letters to Francis Paynter, were published by Walter Moyle in 1726.[12] Of greater biographical value were twelve letters from Lhwyd to Tonkin included by William Pryce in his *Archæologia Cornu-Britannica* (1790). Henry Rowlands published five important letters from Lhwyd, which have not survived in manuscript, in his *Mona Antiqua Restaurata* (1723).[13] In the early nineteenth century, Welsh antiquarian periodicals began to publish Lhwyd's correspondence, in the *Cambrian Register* (1795–1818), the *Cambro-Briton* (1821–2) and the *Cambrian Quarterly Magazine* (1829–33). After transcribing some Lhwyd letters from a copy of BL, Harleian MS 2288 in 1825, the diligent antiquary Angharad Llwyd (1780–1866) told Gwallter Mechain (Walter Davies, 1761–1849):

> Having about 70 of Ed. Lhwyd's original letters to Mr. Mostyn, Dr. Foulks and Mr. Lloyd of Ruthin ... I should like your opinion in regard to which is the best mode of giving them to the world and whether I had better attach to them a life of his publ. about 40 years ago [Huddesford's 'Memoirs'] ... I will give you a short one by way of specimen, two of his Parochial queries are among them ... I should like to see more of his Letters[14]

Much of Lhwyd's correspondence, mainly letters preserved in the Hengwrt manuscripts, was published in *Archaeologia Cambrensis* from 1847 onwards.

Gunther maintained that the 'fate of Lhwyd's Scientific Collections' was 'not less distressing than that of his manuscripts' (G. p. 558). The fragile pressed plants may well have survived better than the apparently indestructible fossils. Lhwyd apparently did not have his own *hortus siccus*, but specimens he had collected can be found in several collections.[15] His lasting contribution to botany was that he provided the first British record of seventeen plants,[16] as well as the first Welsh record of very many mountain plants. The editor of the third edition of Ray's *Synopsis* (1724), J. J. Dillenius (1684–1747), noted that 'observations' by Lhwyd had been taken from his own notes.[17]

Lhwyd gave his collection of fossils to the Museum in 1708,[18] though the donation was not entered in the Book of Benefactors until after his death. Unlike Woodward, he made no arrangements for its custody and maintenance. When Uffenbach was finally allowed to view the collection on 13 September 1710 (it was kept locked, Parry having the only key), he was impressed:

> There is a splendid quantity and variety of these stones, such as I have never in all my life seen together before ... following the description in the book [the *Ichnographia*] they are faultlessly arranged according to class and species ... the larger stones are to be seen uncovered in the big drawers, the smaller ones in round boxes according to size. Those placed thus together are numbered, so that one can find them in the catalogue, and also that they may not get mixed up with each other.[19]

Forty-five years later, when William Huddesford became Keeper, he found the fossil collections in disarray. He told Emanuel Mendes da Costa (1717–91) in 1758 that there had been 'great depredations made of late years', since those in charge 'could not prevent the fossils, in the confusion they were in, from being now and then plundered'.[20] There was some debate about how best to deal with missing items, Mendes da Costa advising Huddesford that to preserve the integrity of the collection, replacements should be identified as such and irreplaceable losses noted.[21] Huddesford's work was undone by his successors, as the fossil collection was dispersed for teaching purposes. In 1925 Gunther believed that only two specimens, from some two thousand, could be identified,[22] but sixteen in all have now been found.[23] Other collections with links to Lhwyd were similarly unfortunate. Those of Sibbald, to whom Lhwyd had given specimens,[24] and Andrew Balfour at Edinburgh were dispersed,[25] and ill-judged disposals at the British Museum in the early nineteenth century reduced Sloane's splendid collection of 4,000 fossils to some 150 identifiable today.[26] The losses are especially unfortunate since several had been presented to Sloane by Lhwyd as particularly fine specimens.[27] There was one fortunate survival in Oxford, for during the later 1920s Gunther discovered a cabinet in the library of Oriel College containing a number of fossils, each wrapped in a paper inscribed in Lhwyd's hand, noting names and find locations and numbers linking them to the *Ichnographia*.[28] This was the collection of Lhwyd's close friend, Richard Dyer (1651–1730),[29] now on loan to the Oxford History of Science Museum.[30] Some appear to have been specimens used for the plates of the *Ichnographia*.

Huddesford published a revised edition of 300 copies of the *Lithophylacii Britannici Ichnographia* at his own expense in 1760.[31] Despite his admiration for Lhwyd, Huddesford realised the need for revision, telling Mendes da Costa in 1758:

> The descriptions are not distinct, the new coined terms, &c. render it very difficult to be understood … Mr Lhuyd's knowledge I dreaded to question – and of his carelessness I had no suspicion; but now I am sensible that he was sometimes deficient in the former and very often guilty of the latter.[32]

Although Huddesford was correct in saying 'erratas abound without number both in the text and plates', this took no account of the problems (noted in Chapter 6) that Lhwyd had experienced in seeing the book through the press, and his frustration that his errata had been omitted. Errors were now corrected, sometimes using Lhwyd's own errata. The new edition reused (with some retouching to improve detail) the original copper plates presented by Lister to Lhwyd and preserved at the Museum, and added two new plates. It included Lhwyd's essay, *Praelectio de stellis marinis Oceani Britannici*, previously published in Leipzig in 1733. Huddesford also prepared a biographical essay, 'Memoirs of the Life of Edward Llwyd', which was published in Nicholas Owen's *British Remains* in 1777. Perhaps initially intended for the *Ichnographia*, it became too bulky for inclusion there, and Huddesford became interested in producing an even fuller life. The 'Memoirs' is of great importance as the first attempt to compile a reasonably full account of Lhwyd's life based on original sources.[33]

Some of Lhwyd's assistants and correspondents continued their researches. Henry Rowlands composed three substantial works during Lhwyd's lifetime. The *Idea Agriculturæ* was written in 1704, but not published until 1764. *An Account of the Origin and Formation of Fossil-shells* appeared anonymously in 1705; its attack on Woodward and development of Lhwyd's seminal theory, as well as a reference to the author showing shells to Lhwyd, argue strongly in favour of Rowlands being the author. His description of the Menai commote, *Antiquitates Parochiales* (1710), intended to form part of a larger survey, shows him to be a true pupil of Lhwyd.[34] The virtues of Rowlands's *Mona Antiqua Restaurata*, part of which was composed during Lhwyd's lifetime, have been obscured by its druidic obsession. Amongst the many works published by Moses Williams, his *Cofrestr o'r Holl Lyfrau Printjedig* (1717), the first published bibliography of printed books in Welsh, and his *Repertorium Poeticum* (1726), a first-line index of Welsh poetry,[35] were in the spirit of Lhwyd's work. Humphrey Foulkes drew on the Lhwyd material at Beechwood for his unpublished volume 'The Modern Antiquity of Wales'.[36] Since Foulkes did not take up Lhwyd's challenge to produce an expanded edition of the *Dictionarium Duplex*, Lhwyd included as Chapter 5 of the *Glossography* (pp. 212–21) a list of 'Some Welch Words Omitted In Dr. Davies's Dictionary'. Foulkes's friend, Thomas Lloyd (1673?–1734) of Plas Power, added many thousands of words to his own copy of the *Dictionarium Duplex*, but these were never published. As noted in Chapter 7, Lhwyd's friend William Gambold had compiled a dictionary by 1722 but was unable to finance its publication. The manuscript passed to John Walters (1721–97), who used it as the basis of his own dictionary published from 1770 onwards.[37]

The decades following Lhwyd's death have been viewed as a time of stagnation, if not decline, in earth studies and natural history in Britain, though Porter has maintained that this was to some extent offset by a growing interest in these subjects by the provincial middle classes.[38] As new explanatory frameworks, most notably stratigraphy, were developed, much of Lhwyd's work came to appear outdated. The adoption of Linnaean taxonomy with its binomial nomenclature from the 1750s onwards made much of Lhwyd's work less accessible to later researchers.

The declining number of references to Lhwyd in biographical dictionaries throughout the nineteenth century is a sign of waning scientific
interest in him. Lhwyd's interest in starfish, however, prompted Edward
Forbes to name a genus of starfish '*Luidia*' to commemorate 'one of the
earliest and most judicious observers of the British Radiata' (G, p. 447).
Following the revision of the nomenclature of *Lloydia serotina* this is
now the only scientific attribution commemorating Lhwyd.

Antiquarianism became less rigorous during the eighteenth century as it became increasingly bound up with the Picturesque and
Druidomania.[39] William Stukeley (1687–1765) epitomises this decline;
as Piggott has shown, he was a sound field archaeologist during the
1720s before succumbing to Druidomania via 'Celtic Temples' in the
early 1730s.[40] Since Lhwyd, like his contemporaries, assigned far too
short a time-scale for the pre-Roman past, erroneous interpretations
of place-names led him to believe that the Neolithic and Bronze Age
monuments found in every Celtic country could be druidic remains and
as such were evidence of a cultural and religious unity. Some antiquarians, notably Toland and Stukeley, seized upon Lhwyd's work to develop
elaborate accounts of the druids. Their fantasies and Pezron's colourful
national 'history', available in an English translation as *The Antiquities
of Nations* (1706), appealed to readers and to artists.[41] Iolo Morganwg
(Edward Williams, 1747–1826) built on these materials to construct
his fantasy of Wales as the true home of the druids, whose traditions
had been continued to his day by the bards.[42]

During Lhwyd's lifetime his activities had impressed some Welsh
people who might not have fully understood his linguistic work, men
such as Thomas Dafydd (see Chapter 10) and the country poet Robin
Rhagad (Robert Humphreys, d. 1718), whose commendatory *englynion*
were published in the *Archæologia*.[43] Lhwyd's travels and correspondence
brought him into contact with local antiquaries, scribes and manuscript
collectors such as Iaco ab Dewi (James Davies, 1648–1722), who composed two rather poor *englynion* praising Lhwyd,[44] and Dafydd Manuel
(David Manuel, 1624?–1726), who appended three *englynion* praising Lhwyd to his answers to Lhwyd's questionnaire.[45] Lhwyd was still
remembered by the next generation, notably by John Morgan, Matchin

(1688?–1733), who published a series of elegiac *englynion* expressing regret that Lhwyd had not completed his work.[46] These are noteworthy for referring to Lhwyd's geological and antiquarian studies, and also – perhaps ominously – for praising him for finding 'all the teaching of the druids'. Lhwyd thus became an iconic figure, providing the scholarly foundation for the myth that the Welsh were the original settlers of Britain and that Welsh was its native language, a myth that restored confidence in a culture and language at a time when its status was rapidly declining.

Nowhere was this myth more powerful than amongst the Welsh community in London. Lewis Morris (1701–65) found much of interest in Lhwyd's work. Many of Lhwyd's research interests appeared in the 'General Heads' of subjects to be considered by the Cymmrodorion Society, which Morris included in its 1755 Constitution.[47] His *Celtic Remains*, composed about 1757 but not published during his lifetime,[48] was a conscious imitation of Lhwyd's projected geographical and historical dictionary; as Morris told Evan Evans (1731–88) in July 1751: 'I am now ... collecting the names of these famous men & women mentiond by our poets, as Mr Ed. Lhwyd once intended.'[49] To a considerable extent, Evans's own work in transcribing old Welsh poetry and eventually publishing English translations of it in his *Some Specimens of the Poetry of the Antient Welsh Bards* (1764) had been inspired by Lhwyd's researches and drew upon his listing of Welsh manuscripts. The *Specimens* also embodied another of Lhwyd's aims by demonstrating the antiquity of the Welsh poetic tradition. In the 1790s Iolo Morganwg and Gwallter Mechain considered publishing the *Celtic Remains*.[50] Gwallter Mechain, one of the *hen bersoniaid llengar* (old literary clerics), knew a considerable amount about Lhwyd's manuscripts and copied some of his letters.[51] When he invited Iolo Morganwg to come to Oxford to 'rummage Edward Llwyd's papers', he mentioned Lhwyd's travel journal, which contained, 'the names of several parishes in north Wales, their townships, rivers, hills, crosses, inscriptions, saints' wells, natural curiosities &cc., chiefly about Oswestry and Flintshire'. This, probably a copy of some responses to the *Parochial Queries*,[52] prompted him to note: 'I may add more parishes to it some time or other and have it bound up.'[53] He published in periodicals from 1795 onwards descriptions of several mid-Wales and border parishes which were in the tradition of the *Parochial Queries*.[54] Accounts of Fishguard and Llanrug also appeared in the *Cambrian Register* in 1795, the latter set out in numbered sections as if responding to a questionnaire, perhaps the detailed queries for 'a parochial history of Great Britain' drawn up by James Theobald (1688–1759) for the Society of Antiquaries, which was published in the *Gentleman's Magazine* in 1755. Histories of Llanerfyl, Llangadfan and Garth Beibo by William Jones (1726–95) were post-humously published in the *Cambrian Register* in 1796.

The greatest contribution of the London Welsh societies was their publication of Welsh texts – *Barddoniaeth Dafydd ab Gwilym* (1789); the *Myvyrian Archaiology* (1801–7), its title a deliberate evocation of Lhwyd's work; *Gwaith Lewis Glyn Cothi* (1837, 1839) – and, more directly relevant to the Lhwydian inheritance, their support of periodicals. The earliest of these, the *Cambrian Register*, was divided into sections containing historical and literary texts, laws, poetry, parish descriptions, lists of collections of manuscripts and the correspondence of past scholars with a selection of their papers. The content of these periodicals thus reflected many of Lhwyd's interests but their uncritical inclusion of Iolo Morganwg's forgeries fell far short of Lhwyd's scholarly standards.[55]

Victor Tourneur (1878–1967) in his comprehensive survey of the history of Celtic studies, *Esquisse d'une histoire des Études celtiques* (Liège, 1905), viewed Lhwyd, whose work had been generally ignored throughout the eighteenth century, as the founder of comparative Celtic linguistics.[56] Leibniz, admittedly, had realised that Irish was important for the study of Celtic languages and had praised the good start, '*egregiê facere cœpit*', made by Lhwyd,[57] and it has been claimed that J. G. Eckhardt depended totally on the *Archæologia* in his discussion of Old Irish.[58] In Wales the detailed discussion of phonetic correspondences of the *Glossography* gave way to the speculative theorising of Rowland Jones (1722?–74) in *The Origin of Languages and Nations* (1764)[59] and the mechanistic analyses of William Owen Pughe (1759–1835).[60] In Ireland, Charles Vallancey (1725?–1812) made very selective use of the

Archæologia in *A Grammar of the Iberno-Celtic or Irish Language* (Dublin, 1773) to support his thesis (p. 123) that since Welsh was 'a corrupted Celtic ... the Irish dialect is the most pure'. In early nineteenth-century discussions of how Celtic languages could be included in the 'new' Indo-European family formulated in 1786 by William Jones (1745–94), one of the few who showed a first-hand knowledge of the *Archæologia* was James Cowles Prichard (1786–1848), who made considerable use of Lhwyd's phonetic correspondences in his *The Eastern Origin of the Celtic Nations* (London, 1831). In 1837, Adolphe Pictet (1799–1875) showed no familiarity with the *Archæologia* in his *De l'affinité des langues celtiques avec la Sanscrit*, referring to Lhwyd only as a source of Cornish vocabulary. J. K. Zeuss (1806–56) referred to Lhwyd in the preface to his pioneering *Grammatica Celtica* (Leipzig, 1853) (p. x), but seems to have made little use of his work. Charles H. H. Wright (1836–1909) in *A Grammar of the Modern Irish Language* (London, 1860) considered Lhwyd to be a century and a half ahead of his time. In mid-nineteenth-century Wales, Thomas Stephens (1821–75) understood the significance of the *Archæologia*, but his was a lonely, unheeded voice.[61] John Rhŷs (1840–1915) expressed his admiration of the *Archæologia* in his *Lectures on Welsh Philology* (1877); in 1896 he delivered a speech in Oswestry urging the erection of a memorial to Lhwyd there.[62] His most eminent pupil, John Morris-Jones (1864–1929), claimed that after he had first encountered Lhwyd's name in Rhŷs's *Lectures*, he set about discovering what he could about 'the great linguist'; he chose 'Edward Lhwyd' as the subject for his inaugural lecture in Bangor in 1893 because the name was so unfamiliar.[63]

By the beginning of the twentieth century Edward Lhwyd was beginning to regain his place as a pioneer in natural history, linguistics and the study of antiquities. It was Richard Ellis (1865–1928), who did most to introduce Lhwyd to an early twentieth-century readership. He founded the Edward Lhuyd Society in Oxford in 1903, and in 1908 published 'Some incidents in the life of Edward Lhuyd',[64] the first substantial biographical study since Huddesford, a work based on a detailed knowledge of the original sources. Unfortunately he then became so obsessed with organising his material that he failed to write a full biography.[65]

R. T. Gunther (1869–1940) provided a wealth of materials for a study of Lhwyd with the posthumous publication in 1945 of his *Life and Letters of Edward Lhwyd*. This made available much of Lhwyd's correspondence, the most important source for tracing the development of his ideas, but was more a quarry than a biography. Ellis's essay thus remained the only modern life of Lhwyd until the publication in 1971 of F. V. Emery's masterly concise study, *Edward Lhuyd F.R.S., 1660–1709*.

Since Lhwyd's realisation that the Celtic languages had a common origin underlies the creation of the Centre for Advanced Welsh and Celtic Studies in Aberystwyth, it is appropriate that a bust of Lhwyd by John Meirion Morris, based on the portrait in the Ashmolean Book of Benefactors, stands in front of the building opened in 1993. The Centre has contributed a new edition of Lhwyd's correspondence as part of an international project, *Early Modern Letters Online* (*EMLO*), and a three-year project (2013–16), Atlantic Europe in the Metal Ages (AEMA), has employed new evidence from a wide range of sources to reconsider the received view of Celtic origins.

Lhwyd is also commemorated today in other ways. There are Edward Lhwyd buildings on the Aberystwyth University Campus and on that of the Owain Glyndŵr University in Wrexham, and a water sculpture inspired by Lhwyd's work on fossils can be found in the National Botanic Garden of Wales. *Canolfan Edward Llwyd* (The Edward Lhwyd Centre) of the Welsh universities is a virtual entity which focuses on Welsh-language resource development. Lhwyd's work as a naturalist inspired the present Cymdeithas Edward Llwyd,[66] founded in 1978 to enable its members to appreciate the natural world and learn about it through the medium of Welsh.[67] The Society realised Richard Ellis's unfulfilled aim of setting up two memorial plaques in Oxford to Lhwyd, one in the chapel of Jesus College (2005) and the other in the church of St Michael (2006). Gardd Edward Lhuyd (the Edward Lhuyd Garden) at Oriel Eryri in Llanberis, Gwynedd, was opened in 1984, but it is currently intended to redevelop the site.

In 1967 Glyn Daniel deplored 'the obscurity into which Lhwyd's archaeological work had fallen ... most historians of archaeology have been scandalously unaware of [Lhwyd's] contribution to the development of their subject.'[68] Daniel concentrated on Lhwyd's contribution to prehistoric archaeology, while recent work has emphasised his important contribution to the study of the post-Roman period,[69] particularly the value of his detailed descriptions of monuments and inscriptions subsequently destroyed or defaced.[70] Scholars such as Graham Parry have examined his antiquarian activities,[71] and his role in the vexed question of 'Celtic' archaeology has come under new scrutiny. Perhaps the most pressing need is for a new critical edition of the *Parochialia* which would integrate material found since Morris's edition. Digitised versions of the drawings in BL, Stowe MSS 1023 and 1024 (ideally with a critical commentary settling questions of attribution) would also be of great value. Meanwhile new uses are being found for the *Parochialia*, most recently as a source for place-name studies.[72]

Although, as with the ichthyosaur vertebrae discussed in Chapter 5, Lhwyd could not have realised the true nature of many of his fossil finds, his meticulous recording and description of specimens and the generally high quality of the depictions in the *Ichnographia* have enabled later palaeontologists to make effective use of his work. In 1753 Mendes da Costa used Lhwyd's depiction of a trilobite to identify a specimen from Dudley.[73] In recent years Lhwyd has been claimed, on the basis of specimen 1570 of the *Ichnographia*, to have been the first to publish a description of a coprolite.[74] He is similarly credited with the first description of a theropod tooth, perhaps from a *Megalosaurus*, in specimen 1328 of the *Ichnographia*.[75]

After considering Lhwyd's career, the man himself remains something of an enigma. Hearne, a close friend, described him as 'a person of singular Modesty, good Nature, & uncommon Industry. He lives a retir'd life … is not at all ambitious of Preferment or Honour, & wt he does is purely out of Love to ye Good of learning & his Country.'[76] Lhwyd was convivial but, as he himself admitted, could be short-tempered. A good hater, he could sometimes be unaccountably forgiving. Hearne considered him to be

a Man of *indefatigable Industry* and of an *enterprizing* and *daring Genius* whom no *Difficulties* or *Hardships* could deterr or frighten from prosecuting his *worthy* and *laudable Designs*; and therefore as nothing *uncommon* and *fit to be noted* could escape his *Inquiry*, so he would never rest satisfied 'till he came to a View of it himself.[77]

In all his work, Lhwyd insisted on personal observation whenever possible and disliked poorly grounded theorising; as he told John Lloyd in 1695,

> I think there are, and always were, too many writers unwilling to acknowledge their ignorance, & therefore loath to give us the bare Relation of several things remarkable, because they knew not the causes of them &c. For my Part when I think I know causes I adde them; and when I do'nt the Reader will have the pleasure of discovering them. (G, p. 273)

NOTES

Introduction

1. Brynley F. Roberts, 'Edward Lhuyd y Cymro', *NLWJ*, 24/1 (1985), 63–83.
2. *Texts*, p. 55.
3. See Brynley F. Roberts, 'Lloyd – Lhuyd – Lhwyd', *Y Traethodydd*, 151 (1996), 180–3.
4. A letter from Ray 7 November 1692 suggests that Lhwyd had recently asked Ray to use this form in addressing letters to him, RFC, p. 232.
5. Roberts, 'Cymraeg', pp. 218–19, 221–2.
6. Brynley F. Roberts, 'Barddoniaeth Edward Lhwyd', *BBCS* 28/1 (1976), 31–44 (pp. 34–6).
7. 'A Letter from Mr. Edward Lhuyd ... giving an Account of a Book, Entituled, ΟΥΡΕΣΙΦΟΙΤΕΣ Helveticus', *PT*, 26 (1708–9), 143–67 (p. 162).
8. *Camden's Wales*, p. 68.
9. Leoni, p. 197.
10. Elizabeth Yale, *Sociable Knowledge: Natural History and the Nation in Early Modern Britain* (Philadelphia, 2016), p. 41.
11. Poole, pp. xii–xiv.
12. A valuable recent discussion is Michael Hunter, *The Decline of Magic: Britain in the Enlightenment* (New Haven and London, 2020).
13. *RFC*, p. 270.
14. Bodleian, MS Ashmole 1815, fol. 81 (*EMLO* image).

Chapter 1

1. Lhwyd's translation of the Irish Preface to his Irish–English dictionary, *Texts*, p. 199. Elsewhere, Lhwyd provided the Welsh forms of some of these names, 'Elidyr Lydanwyn, the son of Meirchion, the son of Ceneu', Roberts, 'Memoirs', pp. 73, 79.
2. A. H. Dodd, 'The Early Days of Edward Lhuyd', *NLWJ*, 6/3 (1950), 305–6 (p. 305).
3. Listed on a flyleaf of Lhwyd's copy of Pezron's *Antiquité ... des Celtes* (1703), Oxford University, Ashmole F 20.
4. See Anthony Ruscoe, *Landed Estates and the Gentry: The Country South of Oswestry* (Ormskirk, 2006), pp. 11, 23; genealogies in *AC* (1867), 118–19; (1885), 59–62; and

J. Y. W. Lloyd, *The History of the Princes, the Lords Marcher, and the Ancient Nobility of Powys Fadog*, 6 vols (London, 1881–7), VI, 357–9.

5. NLW, Peniarth MS 145, pp. 51–4, contains the genealogies and marriage alliances of the Lloyd family of Llanforda and Llwyn-y-maen. In heraldic terms the arms are 'Argent, a double-headed eagle displayed Sable [armed or langued, Gules]', Michael Powell Siddons, *The Development of Welsh Heraldry*, 3 vols (Aberystwyth, 1991–3), II, 380. The tale of how Meurig gained his arms is related in I, 212, where Siddons notes that Gruffudd Hiraethog (d. 1564) was the first to mention this coat.

6. *Victoria History of the Counties of England. A History of Shropshire*, vol. 2, ed. A. T. Gaydon (London, 1973), p. 8. Substantial fines for recusancy imposed on John Lloyd of Llanforda and Richard Lloyd of Llwyn-y-maen in 1588–9 are recorded in *Recusants in the Exchequer Pipe Rolls 1581–1592 ...* ed. Timothy J. McCann, Catholic Record Society Publications (Records Series), vol. 71 (n.p., 1986), pp. 112–13.

7. Ifan ab Owen Edwards, *A Catalogue of Star Chamber Proceedings relating to Wales*, Board of Celtic Studies, University of Wales History and Law Series, no. I (Cardiff, 1929), pp. 221–2.

8. John Pryce-Jones, 'Oswestry Corporation Records – the Bailiffs from Medieval Times to 1673', *Shropshire History and Archaeology, Transactions of the Shropshire Archaeological and Historical Society*, 76 (2001), 30–9 (pp. 35–6); G. Dyfnallt Owen, *Wales in the Reign of James I*, Royal Historical Society Studies in History, 53 (Woodbridge, 1998), pp. 183–4.

9. *Victoria History of the Counties of England: A History of Shropshire*, vol. 3, ed. C. R. Elrington (Oxford, 1979), pp. 88–90.

10. A third storey was added around 1600. Its present appearance dates from an extensive renovation in 1875, John Newman and Nicholas Pevsner, *Shropshire*, The Buildings of England (New Haven and London, 2006), pp. 460–1. Contemporary illustrations show that the roundel with the coat of arms dated 1604 was moved to its present position in 1875, Stanley Leighton, *Sweeney Hall Estate Properties*, pp. 6–9, *http://www.oswestrygenealogy.org.uk/wp-content/uploads/2013/02/leighton. pdf*, accessed 2 October 2020. Historic England considers the style of lettering on the roundel to be late nineteenth to early twentieth century, *https://historicengland. org.uk/listing/the-list/list-entry/1054299*, accessed 25 October 2020.

11. Llinos B. Smith, 'Oswestry', in R. A. Griffiths (ed.), *Boroughs of Mediaeval Wales* (Cardiff, 1978), pp. 218–42 (p. 219).

12. *Gwaith Lewys Glyn Cothi*, ed. Dafydd Johnston (Cardiff, 1995), pp. 455–6.

13. D. J. Bowen, 'Croesoswallt y Beirdd', *Y Traethodydd*, 135/3 (1980), 137–43.

14. *Gwaith Tudur Aled*, ed. T. Gwynn Jones, 2 vols (Cardiff, 1926), I, 235–7.

15. A *cywydd* dated 1561 praises John Llwyd of Llanforda, *Barddoniaeth Wiliam Llŷn a'i Eirlyfr*, ed. J. C. Morrice (Bangor, 1908), pp. 6–8.

16. Daniel Huws, 'Wiliam Llŷn, Rhys Cain a Stryd Wylw', *NLWJ*, 18/1 (1973), 147–8.

17. NLW MS 433, pp. 23–8. I must thank Gruffudd Antur for this reference.

18. For the heraldic work of Rhys and Siôn Cain, see Michael Powell Siddons, *Welsh Pedigree Rolls* (Aberystwyth, 1996).

19. Dafydd Ifans, 'Wiliam Bodwrda (1593–1660)', *NLWJ*, 19/1 (1975), 88–102 (p. 97).
20. The account in Lloyd, VI, 358, is sadly confused. For Caulfeild's appointment, see W. R. Williams, *The History of the Great Sessions in Wales 1542–1830* ... (Brecknock, 1899), p. 130; a reliable account of his career is provided in his biography in vol. 2 of *The History of Parliament: The House of Commons 1588–1603*, ed. P. W. Hasler, 3 vols (London, 2006); see *https://www.historyofparliamentonline.org/volume/1558-1603/member/calfield-%28calfehill-caulfeild%29-george-1603*, accessed 28 October 2020.
21. *Shropshire Parish Registers ... Oswestry*, p. 301.
22. *Alum. Oxon.*
23. He was buried 16 March 1634, *Shropshire Parish Registers*, vol. IV: *The Register of Oswestry*, vol. I (n.p., 1909), p. 504.
24. NLW, Peniarth MS 116, pp. 417–20.
25. John Edwards Griffith, *Pedigrees of Anglesey and Carnarvonshire Families with their Collateral Branches in Denbighshire, Merionethshire and other Parts* (Horncastle, 1914), p. 254. A lacuna in the Oswestry parish registers makes it impossible to establish when Lloyd was married.
26. Paul Stamper, *Historic Parks & Gardens of Shropshire* (n.p., 1996), p. 13.
27. 'The true Narrative of Capt. Edward Lloyd's actions & sufferings in & for his Matie's service', *Bye-Gones* (1887), 413–14.
28. Jonathan Worton, 'The Royalist and Parliamentarian war effort in Shropshire during the First and Second English Civil Wars, 1642–1648' (unpublished doctoral thesis, University of Chester, 2015; see *https://chesterrep.openrepository.com/bitstream/handle/10034/612966/Main%20article.pdf?isAllowed=y&sequence=1*, accessed 4 October 2020.
29. George C. Boon, *Cardiganshire Silver and the Aberystwyth Mint in Peace and War* (Cardiff, 1981), pp. 93–105.
30. For Howard, see *ODNB*.
31. Worton, 'The Royalist and Parliamentarian war effort', p. 78.
32. For Capel, see *ODNB*.
33. The tale is related in William Cathrall, *The History of Oswestry* ... (Oswestry, [1855]), pp. 63–5.
34. Terry Bracher and Roger Emmett, *Shropshire in the Civil War* ([Shrewsbury], 2000), p. 29.
35. Worton, 'The Royalist and Parliamentarian war effort', pp. 45–6.
36. *A Briefe Relation of the most remarkeable Feates and Passages of what his most Gracious Majesties Commanders hath done in England* (Waterford, 1644), p. [13].
37. Worton, 'The Royalist and Parliamentarian war effort', p. 210.
38. John Roland Phillips, *Memoirs of the Civil War in Wales and the Marches, 1642–1649*, 2 vols (London, 1874), I, 222–4; II, 174.
39. For Byron, see *ODNB*.
40. For Trevor, see *ODNB*.
41. Phillips, I, 251.
42. S. Sheppard, *The Yeare of Jubile; or England's Releasement* ... (London, 1646), p. 50.

43. Bodleian, MS Ashmole 1825, fols 101–2.

44. It was estimated in 1656 that it would cost at least £700 to rebuild the church and in 1675 a brief for repairs estimated that £1,500 was required, D. R. Thomas, *The History of the Diocese of St. Asaph, General, Cathedral and Parochial*, new enlarged and illustrated edn, 3 vols (Oswestry, 1908–13), III, 52–3.

45. What follows should be read in the light of Jim Bartos, 'The Spirituall Orchard: God, Garden and Landscape in Seventeenth-Century England before the Restoration', *Garden History*, 38/2 (2010), 177–93.

46. For his career, see G. W. Fisher, *Annals of Shrewsbury School*, rev. J. Spencer Hill (London, 1899), pp. 132–72, 187–90.

47. MS Ashmole 1825, fol. 106.

48. *Bye-Gones* (1887), 413; Worton, 'The Royalist and Parliamentarian war effort', p. 83.

49. See, for example, Byron's dismissive comments in Phillips, II, 247.

50. Phillips, II, 384.

51. MS Ashmole 1825, fol. 116.

52. Griffith, p. 198, states he was vicar of Llanllyfni (1623) and Llaniestyn (1633); for the correct date for Llaniestyn, 1631, see Pryce, *Diocese*, p. 177.

53. MS Ashmole 1825, fol. 144.

54. *Bye-Gones* (1887), 413–14.

55. All conversions have been made using the calculator at *https://www. measuringworth.com/calculators/ukcompare/relativevalue.php.*

56. The former gatehouse of Westminster Abbey, often used to hold prisoners charged with treason.

57. John Bradshaw (1602–59) had presided over the trial of Charles I.

58. *Bye-Gones* (1872), 116.

59. *Bye-Gones* (1887), 422–3.

60. This figure cannot be verified since the Shropshire decimation list has not survived. A comprehensive study of the tax suggests that many of the assessment lists were destroyed, J. T. Cliffe, 'The Cromwellian Decimation Tax of 1655, the Assessment Lists', in *Seventeenth-Century Political and Financial Papers*, Camden Miscellany 23, Camden 5th Series, vol. 7 (Cambridge, 1996), pp. 403–92 (p. 408).

61. MS Ashmole 1825, fol. 130.

62. *Shropshire Parish Registers … Oswestry*, p. 567.

63. MS Ashmole 1825, fol. 80.

64. *Shropshire Parish Registers … Oswestry*, p. 607.

Chapter 2

1. *Shropshire Parish Registers, Diocese of St Asaph*, vol. IV: *The Register of Oswestry.*, vol. I (n.p., 1909), p. 504.

2. *Shropshire Parish Registers*, p. 540.

3. *http://www.innertemplearchives.org.uk/detail.asp?id=4688*, accessed 14 January 2020.

4. *Shropshire Parish Registers*, p. 515.

5. Bodleian, MS Ashmole 1825.

6. *Bye-Gones* (1887), 414.

7. NLW, Peniarth MS 357, p. 1 margin has Lloyd's accounts for lodging and pawning his cloak.

8. Roberts, 'Ceredigion', pp. 49–50.

9. Since college buttery books were intended to record the weekly expenditure there of college members, they provide a good (but not infallible) indication of who was in residence.

10. Roberts, 'Ceredigion', pp. 51–2. Lappiton is the local pronunciation of Loppington, near Wem in Shropshire, some thirteen or fourteen miles east of Oswestry; Crew Green, Llandrinio, is to the south-east of Llanforda.

11. Roberts, 'Ceredigion', pp. 53–5.

12. NLW, Sweeney Hall I, III; NLW, Brogyntyn MS 2560, PEF 1/2.

13. In a letter dated March 1675/6 claiming ancestral relief from payment of tolls in Oswestry, Lloyd said: 'you will receaue I presume ere long good store of white herrings from Mrs Pryce out of Cardiganshire', *Bye-Gones* (1887), 417. 'White herrings' were salt-preserved unsmoked herrings.

14. A. H. Dodd, *Studies in Stuart Wales* (Cardiff, 1952), p. 198.

15. Colin Matheson, *Wales and the Sea Fisheries* (Cardiff, 1929) remains of value. J. Geraint Jenkins, *The Inshore Fishermen of Wales* (Cardiff, 1991) includes a chapter on herring fishing.

16. J. Adams, *Index Villaris* (London, 1680), p. 4. For Adams, see *ODNB*.

17. *Bye-Gones* (1895), 140.

18. *Bye-Gones* (1895), 149–50.

19. Paul Stamper, *Historic Parks and Gardens of Shropshire* (n.p., 1996), pp. 13–14.

20. *Bye-Gones* (1887), 414–17; Dodd, pp. 196–7.

21. *Bye-Gones* (1887), 417.

22. Dodd, p. 206; A. H Dodd, 'The early days of Edward Lhuyd', *NLWJ*, 6/3 (1950), 305–6 (p. 305).

23. Brynley F. Roberts, *Edward Lhuyd: The Making of a Scientist* (Cardiff, 1980), p. 10. An utter or outer barrister was one who pleaded outside the bar as opposed to benchers and others who were allowed to plead within. My thanks to Richard Ireland for his help in this matter.

24. For Williams, see *ODNB*.

25. Llanforda Hall was demolished in 1949; for the gardens see *http://search. shropshirehistory.org.uk/collections/getrecord/CCS_MSA4079/*, accessed 17 September 2020.

26. Probably Elizabeth Herbert (*c*.1633–91). She and her husband (both of them prominent Catholics given to injudicious utterances) had been imprisoned during the Popish Plot. While he remained in the Tower until 1684, she was released in May 1680 and may have been seeking a refuge; see *ODNB* under William Herbert (*c*.1626–96).

27. Roberts, *Making*, p. 12.

28. For Fuller, see John Harvey, *Early Nurserymen* (Chichester, 1974), p. 47, and his *Early Gardening Catalogues* (London and Chichester, 1972), p. 16. See also

Malcolm Thick, 'Garden Seeds in England before the late eighteenth century – II, The Trade in Seeds to 1760', *Agricultural History Review*, 38/2 (1990), 105–16, which discusses Fuller's lists of plants.

29. Lloyd was interested in exotic products. In 1677 he complained of being sent rotten mangoes (probably pickled), adding: 'I never have been in ye Indias, but have eaten Mango these 15 years agoe, and know how to distinguish between good and bad.' *Bye-Gones* (1895), 196. Elsewhere he complained about receiving unsatisfactory china oranges and mouldy tobacco.

30. Malcolm Thick, 'The Sale of Produce from Non-commercial Gardens in late medieval and early modern England', *Agricultural History Review*, 66/1 (2018), 1–17.

31. Roderick Floud, *An Economic History of the English Garden* (n.p., 2019), pp. 256–69.

32. *Bye-Gones* (1895), 171.

33. *Bye-Gones* (1895), 183.

34. Paul Stamper, 'Where Samson Pruned Roses', *British Archaeology*, 53 (June 2000), 42.

35. Roberts, *Making*, p. 12.

36. Stamper, *Historic Parks*, p. 14.

37. Oswestry Ramblers, *More Favourite Walks around Oswestry and the Borders* (Llanrhaeadr ym Mochnant, 2011), p. 2.

38. *Bye-Gones* (1895), 170. The reference is to James Shirley (1596–1666), Royalist poet and playwright, see *ODNB*.

39. Nesta Lloyd, 'Meredith Lloyd', *JWBS*, XI/3–4 (1975–6), 133–92 (pp. 169–72).

40. Owen Morris, *The 'Chymic Bookes' of Sir Owen Wynne of Gwydir: An Annotated Catalogue*, Libri Pertinentes, no. 4 (Cambridge, 1997).

41. Dewi Jones, *Tywysyddion Eryri ynghyd â Nodiadau ar Lysieuaeth yr Ardal*, Llyfrau Llafar Gwlad, 25 (Llanrwst, 1993), pp. 6–8; this was the first British record of *Geum rivale*, Pearman, p. 222.

42. Walter E. Houghton, Jr, 'The English Virtuoso in the Seventeenth Century', *Journal of the History of Ideas*, 3/1–2 (1942), 51–73, 190–219 (p. 202).

43. Alan Cook, *Edmond Halley: Charting the Heavens and the Seas* (Oxford, 1998), pp. 115, 205–12.

44. For Gadbury and his project (contemptuously dismissed by Flamsteed, Hooke and Halley), see *ODNB*.

45. For Tyson, see *ODNB*.

46. Microscopes were very expensive; when Pepys bought his for £5 10s. (almost £850 in modern money) he considered that sum to be a 'great price', *The Diary of Samuel Pepys … vol. V: 1664*, ed. Robert Latham and William Matthew (London, 1971), p. 240.

47. NLW, Brogyntyn MS 2566, PEF 1/2, 30 December 1680.

48. For Clodius, see Vera Keller and Leigh T. I. Penman, 'From the Archives of Scientific Diplomacy: Science and the shared Interests of Samuel Hartlib's London and Frederick Clodius's Gottorf', *Isis*, 106/1 (2015), 17–42 (pp. 21–7). My thanks are due to Dr Leigh Penman for generously sharing the fruits of his research on Frederick Clodius.

49. For Hartlib, see *ODNB*.

50. BL, Add. MS 15070, fol. 10.

51. Karin Seeber, 'Jacob Bobart (1596–1680); First Keeper of the Oxford Physic Garden', *Garden History*, 41/2 (2013), 278–84, a critical reappraisal of his origins and career.

52. For Morison see *ODNB*.

53. Harvey, *Early Nurserymen*, p. 44. Morgan is, regrettably, not included in *ODNB* or *DWB*.

54. Ashmolean Library AMS 2, p. 12.

55. Mea Allan, *The Tradescants, their Plants, Gardens and Museum, 1570–1662* (London, 1964), p. 130, Morgan receives plants from Tradescant; p. 175, witnesses his will.

56. Thomas Johnson, *Mercurii botanicii pars altera* (London, 1641), pp. 7–8.

57. Richard Haslam, 'Bodysgallen: A Renaissance Garden Survival?', *Garden History*, 34/1 (2006), 132–44.

58. R. H. Jeffers, 'Edward Morgan and the Westminster Physic Garden', *Proceedings of the Linnean Society London*, 164/2 (1953), 102–33 (p. 102); 168/1–2 (1957), 96–101.

59. J. Burnby, 'Some Early London Physic Gardens', *Pharmaceutical Historian*, 24/4 (1994), 2–8.

60. E. S. De Beer (ed.), *The Diary of John Evelyn* (London, 1959), p. 391.

61. R. T. Gunther, *Early British Botanists and their Gardens, Based on Unpublished Writings of Goodyer, Tradescant, and Others* (Oxford, 1922), p. 353.

62. Charles E. Raven, *English Naturalists from Neckam to Ray: A Study of the Making of the Modern World* (Cambridge, 1947), p. 318; Burnby, p. 6.

63. Edward Morgan to Robert Morison, 3 December 1680, NLW, Brogyntyn MS PEF 1/2, fols 50v–51r; BL, Add. MS 15070, fol. 55.

64. Raven, *Ray*, pp. 181–3.

65. Philip H. Oswald, 'Edward Morgan and his *Hortus Siccus* as a Source of Early Records of Welsh Plants', *Archives of Natural History* (forthcoming). I am very grateful to Arthur Chater for making it possible for me to see this important article.

66. MS Ashmole 1797, 1798, 1799.

67. MS Ashmole 1815, fol. 3v.

68. NLW, Brogyntyn MS 2566, PEF 1/2, fol. 57.

69. Brogyntyn PEF 1/2. fols 50v–51r. The binomial name for *Solidago sarasenica* is *Senecio sarracenicus*, popularly known as the Broad-leaved Ragwort.

70. NLW, Sweeney Hall 7, fol. 53v.

71. BL, Add. MS 15070, pp. 55–57r.

72. A. O. Chater, 'An Unpublished Botanical Notebook of Edward Llwyd', *Botanical Society of the British Isles Welsh Bulletin*, 40 (1984), 4–15.

73. Chater, p. 6.

74. This location has puzzled Shropshire botanists since the only other report of Pennyroyal, 'the only rare plant [Lhwyd] found in the [Oswestry] area', was a find near Llanymynech recorded by John Evans in 1805, Alex Lockton and Sarah Whild, 'The Botanical Exploration of the Oswestry District', *Shropshire Botanical Society Newsletter*, 9 (Autumn 2003), 8–12 (p. 8).

75. BL, Add. MS 15070, pp. 63v–64r.
76. For David Lloyd, a distant kinsman of Lhwyd, see Roberts, 'John Lloyd', pp. 90–2.
77. For Wynne, see *DWB* and *ODNB*.
78. MS Ashmole 1817b, fol. 379 (*EMLO* transcript).
79. There is no record of his death in the Conwy parish register, *The First Volume of the Conway Parish Registers ... 1541 to 1793* (London, 1900).
80. Ash. Lib. AMS 2, p. 12, which says that when Morgan, 'celebrated former Keeper of the botanical gardens at Westminster' heard from Edward Lhwyd that the Ashmolean lacked a *hortus siccus*, he bequeathed his collection to the Museum. The note adds that Morgan had grown almost all the plants himself in the Westminster garden.
81. In July 1677 he was 'endeavouring to get my health' at the fashionable spa at Epsom, *Bye-Gones* (1895), 183.
82. Roberts, *Making*, p. 18.
83. Funeral directions from Lloyd's will, PRO PROB 11/368; burial, *Shropshire Parish Registers, Diocese of St Asaph*, vol. V: *Register of Oswestry*, vols II–III (n.p., 1912), p. 405. At this time, the wealthy and influential were often buried at night by torchlight, the torches being extinguished at the graveside, *The Diary of Samuel Pepys ...* vol. I: *1660*, ed. Robert Latham and William Matthews (London, 1970), p. 249, n. 2.
84. Presumably to buy a mourning ring as was then fashionable, but if so a very generous sum since the usual price for a good mourning ring was a pound; see 'Funerals' in *The Diary of Samuel Pepys*, vol. X: *Companion*, compiled by Robert Latham (London, 1983), pp. 152–3 (p. 153).
85. For Price, a recusant and antiquary, see *DWB*, and Huws, *Repertory*, II.

Chapter 3

1. His adoption of the 'Lhwyd' form of his surname is discussed in the Introduction.
2. Richard R. Oakley, *A History of Oswestry School* (London, [1964]), p. 67.
3. The most recent history is Stephen A. Harris, *Oxford Botanic Garden & Arboretum: A Brief History* (Oxford, 2017).
4. Mavis Batey, *Oxford Gardens: The University's Influence on Garden History* (Amersham, 1982), pp. 52–3.
5. Reproduced in Harris, pp. 12–13.
6. Scott Mandelbrote, 'The Publication and Illustration of Robert Morison's *Plantarum historiae universalis Oxoniensis'*, *Huntington Library Quarterly*, 78/2 (2015), 349–79.
7. Bobart was sometimes known as, and indeed called himself, 'Professor', though never formally appointed. The title was frequently applied in a casual way, Simcock, pp. 33–40.
8. Harris, pp. 26–8. See also 'Bobart's *Hortus Siccus'*, https://herbaria.plants.ox.ac.uk/bol/bobart, accessed 1 June 2021.
9. Mandelbrote, p. 364. Bobart's work was translated into Latin by William Dale of Queen's College, Mandelbrote, p. 366.

10. Uffenbach, p. 55.

11. NLW, Sweeney Hall MS, V, 16.

12. Raven, *Ray*, p. 184.

13. NLW, Peniarth MS 427, reproduced in G, pp. 67–9.

14. 'The Old Ashmolean Museum', in *A History of the County of Oxford*, vol. 3: *The University of Oxford*, ed. H. E. Salter and Mary D. Lobel (London, 1954), pp. 47–9.

15. Arthur MacGregor, *The Ashmolean Museum: A Brief History of the Museum and its Collections* (Oxford and London, 2001), is a reliable summary. R. F. Ovenell, *The Ashmolean Museum 1683–1894* (Oxford, 1986) provides a fuller account. Valuable for the foundation of the Museum is *Tradescant's Rarities: Essays on the Foundation of the Ashmolean Museum, 1683, with a Catalogue of the Surviving Early Collections*, ed. Arthur MacGregor (Oxford, 1983), available at *http://etheses.dur.ac.uk/10281/1/10281_7075.PDF?UkUDh:CyT*, accessed 14 December 2020.

16. Feingold, p. 394.

17. Feingold, p. 397.

18. Feingold, p. 398.

19. Allan Chapman, 'From Alchemy to Airpumps: The Foundations of Oxford Chemistry to 1700', in *Chemistry at Oxford: A History from 1600 to 2005*, ed. Robert J. P. Williams, John S. Rowlinson and Allan Chapman (Cambridge, 2009), pp. 17–51 (pp. 25–30).

20. Royal Society, *https://royalsociety.org/about-us/history/*, accessed 11 September 2020.

21. Michael Hunter, *Science and Society in Restoration England* (Cambridge, 1981), pp. 39–42.

22. J. A. Bennet, S. A Johnston and A. V. Simcock, *Solomon's House in Oxford: New Finds from the First Museum* (Oxford, 2000), p. 13.

23. Ovenell, pp. 14–15.

24. For these early attempts, see Feingold, p. 435.

25. Hunter, pp. 146–7.

26. *Letters of Humphrey Prideaux … to John Ellis*, ed. Edward Maunde Thompson, Camden New Series, 15 (1876), pp. 1–207 (pp. 60–1).

27. Arthur MacGregor, 'The Tradescants as Collectors of Rarities', in *Tradescant's Rarities*, pp. 17–23, provides a brief but reliable account of the collection. His *Ark to Ashmolean: The Story of the Tradescants, Ashmole and the Ashmolean Museum* (Oxford, 1983) is a popular illustrated history.

28. Ken Arnold, *Cabinets for the Curious: Looking Back at Early English Museums* (Aldershot, 2006), Chapter 2, is a methodologically stimulating introduction. Similar English collections are discussed in Arthur McGregor, 'Collectors and Collections of Rarities in the Sixteenth and Seventeenth Centuries', in *Tradescant's Rarities*, pp. 70–97 (pp. 84–90).

29. A brief account of Ashmole's career and interests, summarising his multi-volume study, is provided by C. H. Josten, 'Elias Ashmole, F.R.S. (1617–1692)', *Notes and Records of the Royal Society of London*, 15 (July 1960), 221–30. See also Michael Hunter, *Elias Ashmole 1617–1692: The Founder of the Ashmolean Museum and his World* (Oxford, 1983).

30. Martin Welch, 'The Foundation of the Ashmolean Museum', in *Tradescant's Rarities*, pp. 40–58 (p. 40).

31. Welch, 'Foundation', p. 44.

32. The debate, mainly about Wren's alleged role, is summarised in Ovenell, pp. 18–21; see also Welch, 'Foundation', p. 47.

33. Welch, 'Foundation', p. 49.

34. Bodleian, MS Ashmole 1814, fol. 98 (*EMLO* transcript).

35. Simcock, p. 5 and plate facing p. 10.

36. Ovenell, p. 106.

37. Welch, 'Foundation', p. 49.

38. Ovenell, p. 38.

39. Welch, 'Foundation', p. 46; Ovenell. pp. 13–14.

40. Roos, *Web*, is the standard biography of Lister.

41. Roos, *Web*, pp. 260–73.

42. Gunther, *Biological*, p. 312.

43. Ovenell, pp. 49–53.

44. For Plot, see *ODNB*; there is no modern biography. Lhwyd contributed 'A short account of the author' to the second edition of Plot's *Oxford-shire* (1705). See also 'Robert Plot' in the Oxford University Museum of Natural History Learning More Series at *https//www.oum.ox.ac.uk/learning/htmls/plot.htm*, accessed 10 December 2020, a reliable brief survey with reproductions of plates of fossils; considerably broader in scope than its title suggests is S. Mendyk, 'Robert Plot: Britain's "Genial father of County Natural Histories"', *Notes and Records of the Royal Society of London*, 39/2 (April 1985), 159–77, supplemented by his *'Speculum Britanniae': Regional Study, Antiquarianism, and Science in Britain to 1700* (Toronto and London, 1989), pp. 193–212; Arnold, pp. 45–63, places Plot's museum activities in the context of his 'philosophical histories'.

45. Simcock, p. 2; Ovenell, p. 33.

46. Ovenell, p. 30.

47. The story that Ashmole had intended to provide an endowment but was dissuaded from doing so by John Tillotson (1630–94), who had resented a hostile notice of his lectures on Socinianism by a member of Oxford University, is set out in Welch, 'Foundation', p. 51; Ovenell, p. 27, is sceptical, citing lack of evidence.

48. Bennett, pp. 22–3; pp. 49–57 discuss human and animal bone remains.

49. Simcock, p. 9.

50. For Higgins, see Foster; for his duties, see Ovenell, pp. 39–40, 53–4; for his resignation, Ovenell, p. 59.

51. Ovenell, p. 47.

Chapter 4

1. *RFC*, p. 276, amended from *EMLO*. For Richard Richardson (1663–1741) see *ODNB*.

2. Roos, *Web*, pp. 362–4.

3. Gunther, *Philosophical*, p. 17.

4. Gunther, *Philosophical*, p. 107.

5. Gunther, *Philosophical*, p. 165.

6. Gunther, *Philosophical*, pp. 123, 127.

7. Matriculated at Jesus College 1672, BA 1676, MA 1685, rector of Llansadwrn, Anglesey 1682, Llanenddwyn 1684, Llanfechell 1691, Pryce, *Diocese*, pp. 10, 12.

8. Gunther, *Philosophical*, pp. 48–9.

9. Gunther, *Philosophical*, p. 107.

10. Gunther, *Philosophical*, p. 108.

11. *PT*, 14 (1684), 823–4.

12. Brian H. Davies, 'Edward Lhuyd and Asbestos Paper', *The Quarterly*, 47 (2003), 10–11.

13. Gunther, *Philosophical*, p. 168; Edward Lhwyd, *Lithophylacii Britannici Ichnographia* (London, 1699), Specimen 910, p. 45, illustrated in Table 12.

14. Gunther, *Philosophical*, p. 170; Lhwyd, *Lithophylacii*, p. 49 *Radiolus glandarius*, Specimen 998.

15. Gunther, *Philosophical*, p. 170.

16. Gunther, *Philosophical*, p. 174. Lister had first drawn attention to chirality in shells in a paper printed in *PT* in 1669, Roos, *Web*, pp. 91–2. The Society was interested in such shells, Plot communicating 30 November 1686 details of one which 'had its turnings from ye left to ye right contrary to allmost all Shells hitherto described', Gunther, *Philosophical*, pp. 191–2.

17. Gunther, *Philosophical*, p. 172.

18. For these journeys, see Raven, *Ray*, pp. 112–13, 125–6.

19. Gunther, *Philosophical*, pp. 200–1.

20. See E. C. Nelson, *Sea Beans and Nickar Nuts: A Handbook of Exotic Seeds and Fruits Stranded on Beaches in North-Western Europe*, B.S.B.I. Handbooks for Physical Identification, 10 (London, 2000).

21. Lee Raye, 'Robert Sibbald's *Scotia Illustrata* (1684): A Faunal Baseline for Britain', *Notes and Records*, 72 (2018), 383–405.

22. Gunther, *Dr. Plot*, p. 323; Gunther, *Philosophical*, p. 206.

23. Gunther, *Philosophical*, p. 201.

24. George C. Boon, 'The Llanymynech Roman Imperial Treasure Trove', *The Numismatic Chronicle*, 7th series, 6 (1966), 155–6.

25. Bodleian, MS Ashmole 1816, fol. 41 (*EMLO* transcript).

26. For John Lloyd, see Roberts, 'John Lloyd', pp. 88–93.

27. Roberts, 'John Lloyd', pp. 96–8; William Gibson, 'The Correspondence of Edward Lhuyd and John Wynne', *Flintshire Historical Society Journal*, 32 (1989), 16–24 (pp. 20–1).

28. See Arthur MacGregor, Melanie Mendonça and Julia White, *Manuscript Catalogues of the Early Museum Collections, 1683–1886, Part 1*, BAR International Series, 907 (Oxford, 2000); Arthur MacGregor and Moira Hook, *Manuscript Catalogues of the Early Museum Collections (Part 2): The Vice-Chancellor's Consolidated Catalogue, 1695*, BAR International Series, 1569 (Oxford, 2006).

29. Lhwyd had compiled his catalogue before publication of the first volume of Lister's illustrated catalogue of shells, *Historiæ … Conchyliorum* (1685).

30. Feingold, pp. 438–9.

31. *Elias Ashmole (1617–1692). His Autobiographical and Historical Notes, his Correspondence, and Other Contemporary Sources Relating to his Life and Work*, ed. C. H. Josten, 5 vols (Oxford, 1966), IV, 1728.

32. Gunther, *Philosophical*, p. 183.

33. Bodleian, Ashmolean MS 2.

34. Michael Hunter, *John Aubrey and the Realm of Learning* (London, 1975) is an important reassessment of Aubrey's scholarship.

35. For Wylde, see *John Aubrey, Brief Lives with An Apparatus for the Lives of our English Mathematical Writers*, ed. Kate Bennett, 2 vols (Oxford, 2015), II, 1688–94.

36. The letter dated May 1688 by *EMLO* is misdated since it refers to Lhwyd's Huntingdonshire expedition of May 1691, G, pp. 144–5. The earliest surviving letter from Lhwyd to Lister is dated 19 July 1689, G, pp. 88–90, but he had earlier sent Lister the shells illustrated in the Appendix to Book III of Lister's *Historiae Conchyliorum*, published in 1688, G, pp. 83–5.

37. Sharpe, *Letters*, pp. 94–5.

38. For its possible occurrence in England, see John Lucey, 'Irish Spurge (*Euphorbia hyberna*) in England: Native or Naturalised?', *British and Irish Botany*, 1/3 (2019), 243–9.

39. Gunther, *Philosophical*, pp. 52, 203.

40. See K. Theodore Hoppen, *The Common Scientist in the Seventeenth Century: A Study of the Dublin Philosophical Society 1683–1708* (London, 1970).

41. Sharpe, *Letters*, pp. 82–3.

42. NLW, Peniarth MS 427, fols 91–2 (*EMLO* transcript).

43. Peniarth MS 427, fols 91–2 (*EMLO* transcript).

44. Brynley F. Roberts, 'An Early Edward Lhwyd Glossary', *SC*, 50 (2016), 151–62.

45. John Ray, *Synopsis Methodica Stirpium Britannicarum* (London, 1690), sig. a3r.

46. Lhwyd sent Ray immature ferns he mistakenly considered to be new finds, Raven, *Ray*, p. 247.

47. Ray, *Synopsis*, p. 7. Doody told Lhwyd 16 May 1692: 'I would gladly see a specimen of [the *Corallium minimum*], you truly make it a stone and no vegitable body', MS Ashmole 1814, fol. 410v.

48. J. E. Smith, *The English Flora*, vol. IV, 2nd edn (London, 1830), pp. 330–1. I must thank Arthur Chater for this reference and for his help with this matter.

49. This could not be, as Gunther maintains, Lhwyd's friend Humphrey Foulkes (1673–1737), since he did not gain his DD until 1720. Further confirmation is that in November 1695 Lhwyd told John Lloyd that he was 'not acquainted with him [Dr Foulkes]', G, p. 292. Perhaps the most plausible candidate is Robert Foulkes (*c*.1652–1728), Precentor of Bangor from 1685 onwards, a botanist who subscribed to the *Archæologia*.

50. *RFC*, p. 277. The letter is undated; Gunther suggested 1699, but Raven's date of 1696 is probably to be preferred, not least because in it Ray says he approved of Lhwyd printing his name in his *Parochial Queries* of late 1696.

51. *Transactions of the Horticultural Society of London*, 1 (1812), 328–9.

52. A. O. Chater, '*Lloydia serotina*', *Welsh Bulletin of the Botanical Society of the British Isles*, 38 (July 1983), 3–7; R. Elwyn Hughes, 'Blodeuyn Edward Llwyd', *Y Naturiaethwr*, 13 (1985), 16–18.

53. Clive Stace, *New Flora of the British Isles*, 4th edn (Middlewood Green, 2019), pp. 899–900, where it is given the rarest designation, RRR. See also Lauren Marrinan and Tim Rich, *101 Rare Plants of Wales* (Llanelli, 2019), pp. 156–7. This notes other Lhwyd discoveries now at risk, such as the ferns *Polystichum lonchitis* (pp. 84–5) and *Woodsia ilvensis* (pp. 112–13).

54. Martin J. S. Rudwick, *The Meaning of Fossils: Episodes in the History of Palaeontology*, 2nd edn (Chicago and London, 1976).

55. Gunther, *Dr Plot*, pp. 335–46.

56. For his collection, see A. J. Turner, 'A Forgotten Naturalist of the Seventeenth Century: William Cole of Bristol and his Collections', *ANH*, 11/1 (1982), 27–41, and Arthur MacGregor and A. Turner, 'The Ashmolean Museum', in *The History of the University of Oxford*, V: *The Eighteenth Century*, ed. L. S. Sutherland and L. G. Mitchell (Oxford 1986), pp. 639–58 (p. 647).

57. Although the earliest surviving evidence of the correspondence is Robinson's letter of 11 April 1696, they appear to have been in contact from 1688 onwards.

Chapter 5

1. In *The Naturalist in Britain*, p. 8, D. E. Allen made much of the club, but L. Jessop, 'The Club at the Temple Coffee House – Facts and Supposition', *ANH*, 16/3 (1989), 263–74, cast doubt on much of the evidence. A recent reappraisal by Margaret Riley, 'The Club at the Temple Coffee House Revisited', *ANH*, 33/1 (2006), 90–100, supports the existence of an organised club whose members included Sloane, Petiver, Lister and Robinson; it welcomed specimens from overseas travellers but did not finance expeditions.

2. Bodleian, MS Ashmole 1817a, fol. 142 (*EMLO* image).

3. MS Ashmole 1816, fol. 93 (*EMLO* transcript).

4. MS Ashmole 1817a, fol. 220 (*EMLO* transcript).

5. Born in Llaniestyn, Caernarfonshire, Parry matriculated at Jesus in 1685, BA March 1692. Lhwyd asked Lister in September 1692 to instruct him in 'natural things', G, pp. 166–7.

6. MS Ashmole 1816, fol. 169 (*EMLO* transcript).

7. MS Ashmole 1814, fol. 80 (*EMLO* transcript).

8. Gunther, *Dr. Plot*, pp. 329–32.

9. See *ODNB*.

10. Published as part of George Hickes, *Institutiones ... Moeso-Goethicae*. For Hickes, see *ODNB*.

11. Gunther, *Dr. Plot*, pp. 330–1.

12. *Texts*, pp. 13–15.

13. NLW, MS 1723D, p. 275 (*EMLO* transcript).

14. *RFC*, p. 210.

15. See further Considine, Chapter 12.

16. John Ray, *A Collection* … 2nd edn (London, 1691), sig. A6a.

17. For Nicolson, see *ODNB*. His glossary of Northumbrian words arrived too late to be fully incorporated in the second edition of Ray's *Collection*.

18. MS Ashmole 1816, fol. 454 (*EMLO* transcript, amended from image).

19. MS Ashmole 1816, fol. 459 (*EMLO* transcript, amended from image). For the 'Rules', see *Texts*, pp. 217–31.

20. Ovenell, p. 65.

21. Both were physicians. Christopher Carstensten de Hemmer (*c*.1664–1715) of Ribe (Jutland) subsequently corresponded with Lhwyd. Nic(h)olas Seerup has yet to be identified but may also be from Ribe. He sent Lhwyd a catalogue of books after leaving England but had died by 12 September 1692, MS Ashmole 1815, fol. 182 (*EMLO* transcript).

22. For Beaumont, see *ODNB*.

23. Jonathan Barry, 'John Beaumont: Science, Spirits and the Scale of Nature', in his *Witchcraft and Demonology in South-West England, 1640–1789* (London, 2012), pp. 124–64.

24. Beaumont's prospectus is reproduced in Gunther, *Dr. Plot*, pp. 275–8.

25. For Petiver see *ODNB*.

26. Anna Marie Roos, 'Only meer love to Learning: A Rediscovered Travel Diary of Naturalist and Collector James Petiver (*c*.1665–1718)', *Journal of the History of Collections*, 29 (2017), 381–94 (p. 386).

27. For the Faringdon Sand Formation, see *https://www.bgs.ac.uk/lexicon/lexicon. cfm?pub=FDNS*, accessed 19 September 2020; the website mentions old gravel pits where the formation is exposed.

28. For Madox, see *ODNB*.

29. Charleton was an alias employed by William Courten, probably as a protection against his very numerous creditors. He set up a private museum in the Temple in 1684, which, bequeathed to Sloane, became part of the original collection of the British Museum. He and Lhwyd had previously exchanged specimens.

30. See further Ovenell, p. 66.

31. Bodleian, MS Ballard 18, fols 3–4 (*EMLO* transcript).

32. *EMLO*, *http://emlo-portal.bodleian.ox.ac.uk/collections/?catalogue=hadriaan-beverland*, accessed 26 September 2020.

33. Ovenell, p. 68.

34. Hearne, I, 253.

35. Ovenell, pp. 68–9.

36. BL, Sloane MS 3962, fols. 293–4 (*EMLO* transcript).

37. Ovenell, p. 69.

38. Ovenell, p. 75.

39. Bodleian, MS Lister 36, fol. 36 (*EMLO* transcript).

40. For Richard Napier (1559–1634), a friend of Simon Forman and John Dee, see *ODNB*.

41. NLW, Peniarth MS 427, fol. 93 (*EMLO* transcript).

42. It has not been possible to identify which of the many William Joneses at Oxford he might have been, Roberts, 'Protégés', p. 47, n. 87.
43. BL, Sloane MS 4067, fol. 25 (*EMLO* transcript).
44. Hearne, VI, 123.
45. Ovenell, p. 91.
46. Richard Ovenden, 'The Learned Press: Printing for the University', in *The History of Oxford University Press*, vol. I: *Beginnings to 1780*, ed. Ian Gadd (Oxford, 2013), pp. 279–92 (pp. 287–8).
47. Hock Cliff, Fretherne SSSI, remains a classic site for collecting Jurassic Blue Lias fossils, *https://ukfossils.co.uk/2010/03/04/hock-cliff/*, accessed 19 September 2020; Tites Point at Purton Passage SSSI is a Regionally Important Geological Site for Upper Ludlow Silurian fossils, *https://ukfossils.co.uk/2013/07/11/tites-point/*, accessed 19 September 2020.
48. For this important fossil location, see N. J. Snelling, *Rocks and Fossils* at *http://www.faringdon.org/uploads/1/4/7/6/14765418/faringdons_fossil_information_njsnelling.pdf*
49. This site, now forested, may well be the 'Oyster Hill' marked on OS maps to the north of Headley, noted in S. Lewis, *Topographical Dictionary of England*, 7th edn (London, 1848) as 'containing fossils of that shell-fish'. The OS *Map of Surrey 1-10560* (1871–82), Sheet 026, *https://www.british-history.ac.uk/os-1-to-10560/surrey/026*, accessed 20 September 2020, notes: 'Fossil Oysters found here at present.'
50. B. C. Moon and A. M. Kirton, *Ichthyosaurs of the British Middle and Upper Jurassic. Part 1-Ophthalmosaurus*, Monograph of the Palaeontographical Society, 170 (647) (London, 2016), pp. 1–84 (p. 16).

Chapter 6

1. See Joseph M. Levine, *Dr. Woodward's Shield* (Berkeley, 1977).
2. Bodleian, MS Ashmole 1817b, fol. 358 (*EMLO* transcript).
3. MS Ashmole 1814, fol. 303 (*EMLO* transcript).
4. MS Ashmole 1817b, fols 348–9 (*EMLO* transcript).
5. The Swedish chemist and mineralogist Erik Odhelius (1661–1704) assumed the name Odelstierna on being knighted in 1698; see *Svenskt biografiskt lexicon* under Odelstierna.
6. Bodleian, MS Eng. hist. c. 11, fol. 109 (*EMLO* transcript).
7. Bodleian, MS Lister 3, fol. 151 (*EMLO* transcript). The letter is undated but the date suggested by *EMLO* is probably correct.
8. MS Ashmole 1817b, fols 353–4 (*EMLO* transcript).
9. MS Ashmole 1817b, fol. 355 (*EMLO* transcript).
10. MS Ashmole 1817b, fols 356–7 (*EMLO* transcript).
11. MS Ashmole 1817b, fols 365–6 (*EMLO* transcript).
12. MS Ashmole 1817b, fols 367–8 (*EMLO* transcript).
13. MS Ashmole 1817b, fols 360–1 (*EMLO* transcript).
14. *Camden's Wales*, p. 90.

15. MS Ashmole 1817b, fol. 373 (*EMLO* transcript).
16. MS Ashmole 1817a, fol. 228 (*EMLO* transcript).
17. MS Lister 36, fols 103–4 (*EMLO* transcript).
18. MS Eng. hist. c. 11, fol. 106 (*EMLO* transcript).
19. MS Eng. hist. c. 11, fol. 115 (*EMLO* transcript).
20. *RFC*, p. 216.
21. Lhwyd's copy, containing his notes and criticisms in Welsh, survives, Ashmole C. 10.
22. For a popular but reliable account of Steno's ideas, see Alan Cutler, *The Seashell on the Mountaintop* (New York, 2003).
23. John Archer (1672/3–1735), nephew by marriage of William Nicolson, who introduced him to Lhwyd, G, p. 125. Having completed his medical studies in Cambridge, Archer moved to London in 1697, Lhwyd recommending him to Lister as 'a person of much candour and ingenuity and my particular friend and patron', G, p. 327.
24. John Morton (1671–1726), curate of Great Oxendon. Inspired by Ray and Lister, he became an enthusiastic collector of fossils.
25. MS Lister 3, fol. 157 (*EMLO* transcript corrected from image). Gunther (p. 278) apparently failed to turn the page and so omitted the last portion of this letter.
26. BL, MS Sloane 4064, fol. 35 (*EMLO* transcript).
27. MS Ashmole 1814, fols 344–5 (*EMLO* transcript).
28. MS Ashmole 1829, fol. 34.
29. Robert Plot, *The Natural History of Oxford-Shire* (Oxford, 1677), p. 121.
30. Matthew R. Goodrum, 'Atomism, Atheism, and the Spontaneous Generation of Human Beings: The Debate over a Natural Origin of the First Humans in Seventeenth-Century Britain', *Journal of the History of Ideas*, 63/2 (2002), 207–24.
31. Rhoda Rappaport, *When Geologists were Historians, 1665–1750* (Ithaca and London, 1997), pp. 192–3.
32. Recorded in Welsh in his copy of Woodward's *Essay*, Bodleian 8° Rawl. 704, see Roberts, 'Cymraeg', pp. 218–19.
33. See G, p. 182, no. 18, for the *Ichthyospondylus*.
34. For his letter to Ray, 30 February [*sic*] 1692, see G, pp. 157–60; for Ray's views, see Raven, *Ray*, pp. 420–37.
35. John Ray, *Miscellaneous Discourses Concerning the Dissolution and Changes of the World …* (London, 1692), p. 44; see also Gordon L. Davies, 'Early British Geomorphology 1578–1705', *Geographical Journal*, 132/2 (1966), 252–62 (p. 258).
36. Dmitry Levitin, 'Halley and the Eternity of the World Revisited', *Notes & Records of the Royal Society*, 67 (2013), 315–29, which distinguishes between arguing for the eternity of the world and arguing for a greatly expanded time-frame.
37. *Camden's Wales*, p. 54.
38. Martin J. S. Rudwick, *Earth's Deep History: How it was Discovered and Why it Matters* (Chicago and London, 2014), pp. 56–8; Poole, p. xvii.
39. MS Lister 36, fol. 22 (*EMLO* transcript).
40. MS Eng. hist. c. 11, fol. 50 (*EMLO* transcript).

41. MS Ashmole 1816, fols 474–5 (*EMLO* transcript).

42. MS Ashmole 1829, fols 16–17 (*EMLO* transcript).

43. MS Ashmole 1816, fol. 196 (*EMLO* transcript).

44. Poole, p. 37.

45. Toland's authorship is asserted in the *ODNB* article on Toland but is scarcely tenable in the light of Rhoda Rappaport, 'Questions of Evidence: An Anonymous Tract Attributed to John Toland', *Journal of the History of Ideas*, 58 (1997), 339–48. Lhwyd, who had known Toland since the spring of 1694, believed that he lacked the requisite knowledge.

46. Tancred Robinson, *A Letter sent to Mr. William Wotton … Concerning Some Late Remarks &c. Written by John Harris* ([London, 1697]), p. [1].

47. MS Lister 3, fol. 135 (*EMLO* transcript).

48. L. P., *Two Essays sent in a Letter from Oxford to a Nobleman in London* (London, 1695), p. 7.

49. L. P., *Two Essays*, p. 8.

50. L. P., *Two Essays*, p. 7.

51. L. P., *Two Essays*, p. 41.

52. Note on p. 2 of Lhwyd's interleaved copy of Woodward's presentation copy of the Essay, Ashmole 8° Rawl. 704.

53. John Harris, *Remarks on some Late Papers relating to the Universal Deluge …* (London, 1697), pp. 113–19.

54. Levine, p. 81.

55. MS Ashmole C.15.

56. For a detailed account and transcriptions, see Brynley F. Roberts, '*Anwir Anwedhys y mae yn i Ysgrivennv Ymma': Rhai o Ymylnodau Edward Lhwyd* (Aberystwyth, 2009), pp. 36–47.

57. MS Eng. hist. c. 11, fol. 94 (*EMLO* transcript).

58. MS Ashmole 1817a, fol. 341 (*EMLO* transcript).

59. MS Ashmole 1829, fol. 146 (*EMLO* transcript).

60. For Mead, see *ODNB*, which notes that he maintained a lifelong enmity towards Woodward.

61. Levine, pp. 15–17.

62. For Williams, see Emery, 'Glamorgan', pp. 62–75.

63. Emery, 'Glamorgan', pp. 67–8.

64. MS Ashmole 1817a, fol. 256 (*EMLO* transcript).

65. *RFC*, p. 264, amended from *EMLO*.

66. Ian Gadd, 'Introduction', in *The History of Oxford University Press*, vol. I: *Beginnings to 1780*, ed. Ian Gadd (Oxford, 2013), pp. 3–28 (pp. 11–12).

67. Vittoria Feola and Scott Mandelbrote, 'The Learned Press: Geography, Science, and Mathematics', in *The History of Oxford University Press*, vol. I: *Beginnings to 1780*, ed. Ian Gadd (Oxford, 2013), pp. 317–57 (p. 350).

68. Scott Mandelbrote, 'The Publication and Illustration of Robert Morison's *Plantarum historiae universalis Oxoniensis*', *Huntington Library Quarterly*, 78/2 (2015), 349–79 (pp. 372–4).

69. Matthew Kilburn, 'The Fell Legacy 1686–1755', in *The History of Oxford University Press*, vol. I: *Beginnings to 1780*, ed. Ian Gadd (Oxford, 2013), pp. 107–37 (pp. 118–21).

70. MS Ashmole 1817b, fol. 64 (*EMLO* transcript).

71. *PT*, 26 (1708), 77–80.

72. MS Eng. hist. c. 11, fol. 87 (*EMLO* transcript).

73. Possibly Louis Feuillée, (1660–1732). An A–Z listing of members is provided at *https://www.academie-sciences.fr/en/Table/Membres/Liste-des-membres-depuis-la-cr eation-de-l-Academie-des-sciences/ les-membres-du-passe-dont-le-nom-commence-par-f. html*, accessed 15 September 2020.

74. MS Eng. hist. c. 11, fol. 88 (*EMLO* transcript).

75. Étienne François Geoffroy (1672–1731), made an Associate of the Académie des Sciences in December 1699, who was also a Fellow of the Royal Society.

76. *Ray Correspondence*, p. 239.

77. *RFC*, p. 204.

78. MS Eng. hist. c. 11, fol. 91 (*EMLO* transcript).

79. MS Eng. hist. c. 11, fol. 76 (*EMLO* transcript).

80. MS Eng. hist. c. 11, fol. 92 (*EMLO* transcript).

81. MS Ashmole 1817b, fols 94–5 (*EMLO* transcript).

82. MS Ashmole 1817b, fols 94–5 (*EMLO* transcript).

83. MS Eng. hist. c. 11, fol. 93 (*EMLO* transcript).

84. BL, Sloane MS 4062, fol. 297 (*EMLO* transcript).

85. Bodleian, Lister L 92.

86. MS Eng. hist. c. 11, fol. 93 (*EMLO* transcript).

87. MS Ashmole 1817b, fol. 96v (*EMLO* transcript).

88. BL, Sloane MS 4062, fol. 309 (*EMLO* transcript).

89. MS Ashmole 1817a, fol. 264 (*EMLO* transcript).

90. Hearne, I, 244.

91. For the book, see Marcus Hellyer, 'The Pocket Museum: Edward Lhwyd's "*Lithophylacium*"', *ANH*, 23 (1966), 43–60.

92. Hellyer, p. 47.

93. Hellyer, p. 54.

94. Lhwyd told Richardson that he had combined 'most of the Observations you have been pleased to communicat your severall Letters, under the Title of an Extract of a Letter from you', BL, Sloane MS 4062, fols 286–7 (*EMLO* transcript).

95. G, nos 191, Nicolson; 192, Archer; 198, Robinson; 200, Ray; 189, Rivinus.

96. R. M. Owens, *Trilobites in Wales* (Cardiff, 1984), pp. 3–4, 8.

97. MS Eng. hist. c. 11, fol. 87 (*EMLO* transcript).

98. For Mostyn, see Chapter 8.

99. For Hales's views, see Goodrum, pp. 214–16.

100. MS Ashmole 1817a, fol. 342 (*EMLO* transcript).

101. MS Ashmole 1817a, fol. 343 (*EMLO* transcript).

102. The latest study is David Deming, 'Edmond Halley's Contributions to Hydrogeology', *Groundwater*, 59/1 (2021), 146–52, available at *https://ngwa. onlinelibrary.wiley.com/doi/full/10.1111/gwat.13059*, accessed 5 June 2021.

103. Bodleian, Ashmole B.7, pp. 177–8.

104. Royal Society, LBO/14, pp. 344–7 (*EMLO* transcript).

105. Explaining physical phenomena by witchcraft was a contentious issue in the 1690s; see Michael Hunter, *The Decline of Magic: Britain in the Enlightenment* (New Haven and London, 2020).

106. BL, Sloane MS 4063, fol. 25 (*EMLO* transcript).

Chapter 7

1. For the remarkable flowering in the study of Anglo-Saxon at Oxford during Lhwyd's lifetime, see David Fairer, 'Anglo-Saxon Studies', in *The History of the University of Oxford*, vol. V: *The Eighteenth Century*, ed. L. S. Sutherland and L. G. Mitchell (Oxford, 1986), pp. 807–29.

2. For Swalle, see Joseph M. Levine, *The Battle of the Books: History and Literature in the Augustan Age* (Ithaca and London, 1991), pp. 327–30, 332, 335.

3. Two older studies of Camden's *Britannia* provide valuable introductions: T. D. Kendrick, '*Britannia*', in his *British Antiquity* (London, 1950), pp. 134–67; and Stuart Piggott, 'William Camden and the *Britannia*', in Piggott, *Ruins*, pp. 33–53. More recent is Parry, pp. 22–48, and Chapter 9 of John Cramsie, *British Travellers and the Encounter with Britain, 1450–1700* (Woodbridge, 2015).

4. Piggott, *Ruins*, p. 43.

5. BL, Add. MS 15070, fols 20–38v. Lhwyd was interested in the section on Brigantia.

6. An important recent study is Thomas Roebuck, 'Edmund Gibson's 1695 *Britannia* and Late-Seventeenth-Century British Antiquarian Scholarship', *Erudition and the Republic of Letters*, 5 (2020), 427–81.

7. Gwyn Walters and Frank Emery, 'Edward Lhuyd, Edmund Gibson, and the Printing of Camden's *Britannia*, 1695', *The Library*, 5th series 32 (1977), 109–37 (p. 127).

8. For Gibson, see *ODNB*.

9. Roebuck, pp. 434–5; for Harrington see *ODNB*.

10. Roebuck, p. 438.

11. Roebuck, p. 456.

12. Roebuck, p. 456; *RFC*, p. 244, where Ray said that Lhwyd was 'abundantly better able to perform that task than my self'.

13. This list was tacitly omitted in *Camden's Wales*.

14. Bodleian, MS Ashmole 1814, fol. 404 (*EMLO* image).

15. *Texts*, pp. 3–4.

16. MS Ashmole 1815, fols 270–2 (*EMLO* transcript).

17. MS Ashmole 1815, fols 273–4 (*EMLO* transcript).

18. *Camden's Wales*, p. 70.

19. MS Ashmole, 1815, fol. 269 (*EMLO* transcript).

20. MS Ashmole 1817b, fol. 381 (*EMLO* transcript).
21. *Camden's Wales*, pp. 70–1.
22. *Camden's Wales*, p. 71.
23. Roberts, 'John Lloyd', p. 103.
24. Roberts, 'John Lloyd', p. 101.
25. Lloyd, 'Correspondence', pp. 31–3.
26. Lloyd, 'Correspondence', p. 38.
27. Lloyd, 'Correspondence', p. 40.
28. Lloyd, 'Correspondence', p. 50.
29. Adam Fox, 'Printed questionnaires, and research networks, 1650–1800', *Historical Journal*, 53 (2010), 593–621; Roberts, 'Folklorist', p. 40.
30. Lhwyd's papers contain several sets of contemporary queries, e.g. Machell, 1677, Molyneux 1682, Plot 1674, 1679; see Roberts, 'Folklorist', p. 39.
31. Roberts, 'John Lloyd', pp. 101–2.
32. MS Ashmole 1816, fols 204–5 (*EMLO* transcript).
33. Roberts, 'John Lloyd', p. 103.
34. MS Ashmole 1815, fol. 26 (*EMLO* transcript).
35. Bangor University Library Archives and Special Collections, Penrhos MS V, 929; I am grateful to the Librarian for permission to publish the document; for Lhwyd's comments on 'ʒn hericy Gwidil', see *Camden's Wales*, p. 80.
36. For Gambold, see DWB; a schoolmaster at Cardigan during the 1690s, rector of Puncheston 1709, he compiled an unpublished Welsh dictionary and published a Grammar in 1727.
37. MS Ashmole 1830, fols 31–2 (*EMLO* transcript amended from image).
38. *Camden's Wales*, pp. 56–7.
39. Roberts, 'John Lloyd', p. 184.
40. G, p. 59. Matriculated at Jesus College 1664/5, BA 1668, MA 1671. Vicar of Llanddewi Felffre, Pembrokeshire, from 1673, Headmaster of Queen Elizabeth grammar school Carmarthen 1672–86, he contributed to Ogilby's road atlas, *Britannia* (1675).
41. MS Ashmole 1817a, fol. 306 (*EMLO* transcript).
42. MS Ashmole 1817a, fols 307–8 (*EMLO* transcript).
43. MS Ashmole 1817a, fols 309–13 (*EMLO* transcript).
44. The following page number references in parentheses are to *Camden's Wales*.
45. MS Ashmole 1817a, fol. 312v (*EMLO* transcript).
46. MS Ashmole 1815, fol. 221 (*EMLO* transcript).
47. MS Ashmole 1815, fol. 265 (*EMLO* transcript).
48. See Nancy Edwards, *A Corpus of Early Medieval Inscribed Stones and Stone Sculpture in Wales*, vol. 3: *North Wales* (Cardiff, 2013).
49. These early medieval graves have been destroyed since Lhwyd's visit; see Richard Hayman and Wendy Horton, *Manod Bach – Y Garnedd: An Archaeological Survey* (n.p., 2014), p. 7.
50. MS Ashmole 1814, fols 379–80 (*EMLO* transcript).
51. MS Ashmole 1814, fol. 381 (*EMLO* transcript).

52. MS Ashmole 1814, fol. 383 (*EMLO* transcript).
53. MS Ashmole 1814, fol. 383 (*EMLO* transcript amended from image).
54. Matriculated at St Alban's Hall, Oxford, 1668/9 aged 22, BA 1672, MA 1675, *DWB*.
55. MS Ashmole 1817b, fols 56–7 (*EMLO* transcript amended from image).
56. MS Ashmole 1817b, fols 297–8 (*EMLO* transcript).
57. MS Ashmole 1817b, fol. 292 (*EMLO* transcript amended from image); the passage as quoted in Emery, 'Glamorgan', omits a line between 'edgewise' and 'of ye Lapis molinaris'.
58. Matriculated at Jesus College, Oxford 1686/7, BA 1690, MA, 1697.
59. MS Ashmole 1815, fol. 247 (*EMLO* transcript).
60. Roberts, 'Carmarthenshire', pp. 36–41.
61. Roberts, 'Protégés', pp. 33–6.
62. Frank Emery, 'A New Account of Snowdonia, 1693, written for Edward Lhuyd', *NLWJ*, 18/4 (1974), 405–17.
63. MS Ashmole 1815, fol. 169 (*EMLO* transcript).
64. For Wilkins, see G. J. Williams, *Traddodiad Llenyddol Morgannwg* (Cardiff, 1948), pp. 161–8; in 1701 Lhwyd copied the manuscript, now Cardiff, MS 3.464; his transcript is in Bodleian, MS Carte 108.
65. MS Ashmole 1815, fol. 171 (*EMLO* transcript).
66. *Camden's Wales*, p. 43.
67. Emery, 'Glamorgan', pp. 65–6.
68. On this version of Geoffrey of Monmouth's chronicle, see Brynley F. Roberts, *Brut Tysilio* (Swansea, 1980).
69. Brynley F. Roberts, '*Anwir Anwedhys y mae yn i Ysgrivennv Ymma': Rhai o Ymylnodau Edward Lhwyd* (Aberystwyth, 2009), p. 22.
70. 'At the same *Kaer Leion* they frequently dig up Roman bricks with this inscription. LEG II. Avg. The letters on these bricks are not *inscrib'd* (as on Stone) but *stamp'd* with some instrument; there being a square cavity or impression in the midst of the Brick, at the bottom whereof the Letters are *rais'd*, and not *inscrib'd*.', *Camden's Wales*, p. 37.
71. MS Ashmole 1817a, fol. 200 (*EMLO* transcript).
72. Roberts, 'Folklorist', p. 37.
73. Roberts, 'Folklorist', p. 45.
74. Walters and Emery, p. 135.
75. *Camden's Wales*, p. 11.
76. Walters and Emery, p. 133.
77. *Camden's Wales*, p. 19, which omits the first paragraph of Lhwyd's letter.
78. Roebuck, pp. 471–3.
79. Roebuck, p. 477.
80. Hearne, I, 217.
81. Hearne, IV, 169.
82. MS Ashmole 1817a, fols 318–19 (*EMLO* image).
83. Parry, p. 347.

Chapter 8

1. Considine, pp. 124–5.
2. Bodleian, MS Ashmole 1816, fol. 254 (*EMLO* transcript).
3. MS Ashmole 1816, fol. 370 (*EMLO* transcript).
4. MS Ashmole 1817b, fols 301–2 (*EMLO* transcript).
5. MS Ashmole 1817b, fol. 303 (*EMLO* transcript).
6. Emery, 'Glamorgan', pp. 72–3, describes Williams's role and analyses the list of subscribers in the *Archæologia*.
7. MS Ashmole 1817b, fol. 311 (*EMLO* transcript).
8. *Texts*, pp. 35–9.
9. Ovenell, pp. 86–7.
10. From St Nicholas, Glamorgan, matriculated at Jesus College in 1693, graduated in 1696; see Roberts, 'Protégés', pp. 37–8.
11. Bodleian, MS Eng. hist. c. 11, fol. 23 (*EMLO* transcript).
12. MS Ashmole 1817b, fols 331–2 (*EMLO* transcript).
13. Ovenell, p. 88.
14. MS Eng. hist. c. 11, fol. 83 (*EMLO* transcript).
15. For the problems caused by the recoinage, see Brodie Waddell, 'The Politics of Economic Distress in the Aftermath of the Glorious Revolution, 1689–1702', *English Historical Review*, 130 (2015), 318–51 (pp. 323–8).
16. MS Ashmole 1817a, fol. 474 (*EMLO* transcript, date amended from 1695 on internal evidence).
17. MS Ashmole 1817a, fols 329–30 (*EMLO* transcript).
18. These are, apparently, the only references to the Tlws.
19. Alan Cook, *Edmond Halley: Charting the Heavens and the Seas* (Oxford, 1998), p. 196.
20. MS Eng. hist. c. 11, fol. 80 (*EMLO* transcript).
21. Nancy Edwards, 'Rethinking the Pillar of Eliseg', *Antiquaries Journal*, 89 (2009), 143–77.
22. Lhwyd's transcript is reproduced in Edwards, 'Rethinking', pp. 158–9.
23. NLW, MS 5262A, fol. 68v.
24. MS Ashmole 1817b, fol. 12 (*EMLO* transcript).
25. Chapter 7, pp. 254–65; see also the section in his Preface.
26. MS Ashmole 1816, fols 199–200 (*EMLO* transcript).
27. *Texts*, pp. 41–7.
28. *Texts*, pp. 46, 42, 44.
29. For printed questionnaires, see Adam Fox, 'Printed questionnaires, research networks, and the discovery of the British Isles, 1650–1800', *Historical Journal*, 53/3 (2010), 593–621; also Roberts, 'Folklorist', p. 40.
30. F. V. Emery, 'A Map of Edward Lhuyd's *Parochial Queries in Order to a Geographical Dictionary, &c., of Wales* (1696)', *THSC* (1958), 41–53 (p. 44).
31. *Texts*, p. 41.
32. Emery, 'Edward Lhuyd', p. 73.

33. MS Ashmole 1817b, fol. 91 (*EMLO* transcript).
34. MS Ashmole 1817b, fol. 84 (*EMLO* transcript).
35. Roberts, 'Protégés', pp. 31–3.
36. Roberts, 'Protégés', pp. 35–6.
37. Roberts, 'Protégés', p. 37; G, p. 301.
38. MS Ashmole 1817b, fol. 393 (*EMLO* transcript); the letter is undated but the context suggests a date of 1696, not 1697 as in *EMLO*.
39. Roberts, 'Protégés', p. 41.
40. MS Ashmole 1817a, fol. 72 (*EMLO* image).
41. Roberts, 'Protégés', p. 41; for agreements with other amanuenses, see pp. 28–9, n. 19.
42. Cardiff, MS 4.120, pp. 5–6 (*EMLO* transcript).
43. MS Ashmole 1817a, fol. 337 (*EMLO* transcript).
44. MS Ashmole 1817b, fol. 84 (*EMLO* transcript).
45. Uffenbach, p. 49.
46. Roberts, 'Books', p. 112.
47. For their fate, see Eiluned Rees and Gwyn Walters, 'The dispersion of the manuscripts of Edward Lhuyd', *WHR*, 7/2 (1974), 148–78.
48. Nancy Edwards, 'Edward Lhuyd and the Origins of Early Medieval Celtic Archaeology', *Antiquaries Journal*, 87 (2007), 165–96 (pp. 182–4).
49. F. J. North, *Coal, and the Coalfields in Wales* (Cardiff, 1926), pp. 99–100.
50. For *Lepidodendron*, see Christopher J. Cleal and Barry A. Thomas, *Plant Fossils of the British Coal Measures*, Palaeontological Association Field Guides to Fossils, 6 (London, 1994), pp. 21–4.
51. Edward Lhwyd, 'A Note concerning an Extraordinary Hail in Monmouthshire', *PT*, 19 (1695–7), pp. 579–80.
52. John Davies, *Antiquæ Linguæ Britannicæ ... Dictionarium Duplex* (London, 1632), sig. 3I5b–3I6a.
53. Printed in Richard Fenton, *Tours in Wales (1804–1813)*, ed. John Fisher (London, 1917), pp. 336–49.
54. For a detailed analysis, see Roberts, 'Carmarthenshire', pp. 31–2.
55. NLW, Peniarth 427, fol. 54 (*EMLO* transcript).
56. For Parry, see Roberts, 'Protégés', pp. 55–6.
57. For Lhwyd and the Pryse family, see Roberts, 'Ceredigion', pp. 60–4 and n. 16.
58. MS Ashmole 1817b, fol. 383 (*EMLO* transcript).
59. *Parochialia*, II, 82; Trefor M. Owen, *Welsh Folk Customs* (Cardiff, 1968), pp. 63–4.
60. *Parochialia*, II, 81–2.
61. Stephen W. Jones, 'Thomas Francis of Montgomery (?1621–1700)', *Journal of the Montgomeryshire Genealogical Society*, 59 (2014), 25–34.
62. NLW, MS 21297E, 11.b (*EMLO* transcript).
63. *Parochialia*, II, 25–83.
64. *Parochialia*, II, 84–108.
65. Edwards, 'Edward Lhuyd', p. 170; the ogam stone from Ystrad, lost until 1957, is reproduced on p. 183.

66. The collection of Sir William Williams of Glasgoed and Llanforda, who had bought the collection of William Maurice; see Huws, *Repertory*, II, s.n. Edward Lhuyd.

67. For the inscription, variously dated between the seventh and the ninth centuries, see Ifor Williams, 'The Towyn Inscribed Stone', in *The Beginnings of Welsh Poetry: Studies by Ifor Williams*, ed. Rachel Bromwich (Cardiff, 1972), pp. 25–40; Nancy Edwards, *A Corpus of Early Medieval Inscribed Stones and Stone Sculpture in Wales*, vol. 3: *North Wales* (Cardiff, 2013), p. 430.

68. See *Camden's Wales*, p. 60; for the 29 March 1699 letter, see E. Gwynne Jones, 'The family papers of Owen and Stanley of Penrhos, Holyhead', *BBCS*, 17/2 (1959), 99–115 (p. 110); and Nancy Edwards, *A Corpus of Early Medieval Inscribed Stones and Stone Sculpture in Wales*, vol. 2: *South-West Wales* (Cardiff, 2007), pp. 150–3.

69. Bodleian, MS Lister 36, 237 (MS Lister CXXIV) (*EMLO* image).

Chapter 9

1. *Texts*, p. 37.

2. Cardiff, MS 4.120, p. 5 (*EMLO* transcript).

3. J. L. Campbell, 'The Tour of Edward Lhuyd in Ireland in 1699 and 1700', *Celtica*, 5 (1960), 218–28 (p. 219).

4. Bodleian, MS Ashmole 1816, fol. 363 (*EMLO* transcript).

5. Nos 193–849 in the *Glossography*, pp. 435–6.

6. Cardiff, MS 4.120, pp. 5–6 (*EMLO* transcript).

7. Alasdair Kennedy, 'In Search of the "True Prospect": Making and Knowing the Giant's Causeway as a Field Site in the Seventeenth Century', *The British Journal for the History of Science*, 41/1 (2008), 19–41 (pp. 21–2).

8. *PT*, 17 (1693), 708–10.

9. *PT*, 18 (1694), 170–82.

10. *PT*, 20 (1698), 209–23.

11. In *PT*, 68 (1778), 1–6, Hamilton pointed out that the basaltic columns of buildings in Cologne, 'very like the basaltes of the Giant's Causeway', were evidence of the volcanic origins of both; see also Evelyn Stokes, 'Volcanic Studies by Members of the Royal Society of London 1665–1780', *Earth Science Journal*, 5/2 (1971), 46–67.

12. Cardiff, MS 4.120, p. 5 (*EMLO* transcript).

13. Christopher Duffin and Jane P. Davidson, 'Geology and the Dark Side', *Proceedings of the Geologists' Association*, 122 (2011), 7–15 (p. 10).

14. Campbell, p. 220.

15. Bronze Age horns, now lost; C. Stephen Briggs, 'Edward Lhuyd's Expeditions through Ulster, 1699 and 1700', in *The Modern Traveller to Our Past: Festschrift in Honour of Ann Hamlin*, ed. Marion Meek (n.p., 2006), pp. 384–93 (p. 391).

16. *Glossography*, p. 436.

17. *ELSH*, p. xiv; Anne O'Sullivan and William O'Sullivan, 'Edward Lhuyd's Collection of Irish Manuscripts', *THSC* (1962), 57–76; Huws, *Repertory*, II, s.n. Lhuyd, Edward.

18. David McGuinness, 'Edward Lhuyd's Contribution to the Study of Irish Megalithic Tombs', *Journal of the Royal Society of Antiquaries of Ireland*, 126 (1996), 62–85.
19. To Robinson, 15 December 1699, G, pp. 421–3, and *PT*, 27 (1699); to Thomas Molyneux, 29 January 1700 and to Henry Rowlands, 12 March 1700, G, pp. 429–30.
20. Reproduced in Geraldine Stout and Matthew Stout, *Newgrange* (Cork, 2008), fig. 66.
21. Dublin, Trinity College 888/2, pp. 312–13 (*EMLO* transcript); cf. a letter from Jones to Lhwyd 6 February 1700, where he speaks of his willingness to go to Newgrange, and of Molyneux being 'desirous of seeing ye draught wh. drawn', MS Ashmole 1815, fol. 295 (*EMLO* transcript).
22. McGuinness, p. 81.
23. Dublin, Trinity College 883/2, p. 291, with additions from Cardiff, MS 4.120, pp. 17–22 (*EMLO* transcript); McGuinness does not appear to have seen these important additions.
24. In 1687, for example, Lhwyd had quoted Sibbald's *Scotia Illustrata* (1684); see Chapter 4.
25. MS Ashmole 1816, fol. 279 (*EMLO* transcript).
26. MS Eng. hist. c. 11, fol. 97 (*EMLO* transcript).
27. Cardiff, MS 4.120, pp. 285–6 (*EMLO* transcript).
28. *ELSH*, p. xii, has a helpful map of Lhwyd's probable itinerary.
29. Royal Society, LBO/14, pp. 355–62 (*EMLO* transcript).
30. G, p. 426; *ELSH*, pp. 6–7.
31. BL, Sloane MS 4063, fol. 25 (*EMLO* transcript).
32. For this monument see *https://portal.historicenvironment.scot/designation/SM1183*, accessed 12 December 2020.
33. See *https://canmore.org.uk/site/50719/the-cat-stane*.
34. First published in 1675 and frequently reprinted, this was a word list in English, Latin and Greek arranged under thirty-two subject headings.
35. See J. L. Campbell, *A Collection of Highland Rites and Customes copied by Edward Lhuyd from the Manuscript of the Rev. James Kirkwood … and annotated by him with the aid of the Rev. John Beaton* (Cambridge, 1975); *ELSH*, pp. xxi, xxvii–xxx, 52–75.
36. Cardiff, MS 4.120, p. 289 (*EMLO* transcript).
37. A recent survey of such use of fossils is Ken McNamara, *Dragon's Teeth and Thunderstones: The Quest for the Meaning of Fossils* (London, 2020).
38. *Camden's Wales*, p. 84.
39. John Davies, *Dictionarium Duplex* (London, 1632).
40. *ELSH*, pp. xvii–xviii.
41. *ELSH*, pp. 4–5; for responses from Beaton and John Frazer, see *ELSH*, pp. 23–36.
42. For a copy (in a different notebook) and discussion of his Scottish Gaelic (Argyll) version, see *ELSH*, pp. 91–217.
43. Roberts, 'Protégés', pp. 49–50.
44. Huddesford, p. 137.

45. See five pages of fines from 1702–5 in Bodleian, MS Ashmole 5/1, and cf. Ashmole 1814, fol. 307.

46. *ELSH*, pp. xxi, 12.

47. Lhwyd noted that, though Beaton was a Highlander, his pronunciation was like that of the Irish of Ireland, Roberts, 'Cymraeg', p. 214; for Beaton, *ELSH* is the essential discussion.

48. MS Ashmole 1815, fol. 286 (*EMLO* transcript). For a draft version of this, with helpful commentary, see Thomas Jones, 'Llythyr gan William Jones at Edward Lhuyd', *BBCS*, 22/3 (1967), 239–43.

49. MS Ashmole 1815, fol. 295 (*EMLO* transcript).

50. Cardiff, MS 4.120, pp. 17–20 (*EMLO* transcript).

51. Campbell, pp. 222–8.

52. Cardiff, MS 4.120, p. 8 (*EMLO* transcript).

53. Cardiff, MS 4.120, pp. 8–10 (*EMLO* transcript).

54. Campbell, p. 227.

55. Máire Lohan, 'Ceremonial Monuments in Moytura, Co. Mayo', *Journal of the Galway Archaeological and Historical Society*, 51 (1999), 77–108 (p. 77).

56. McGuinness, p. 83.

57. Pearman, pp. 47, 62.

58. M. E. Mitchell, 'Irish Botany in the Seventeenth Century', *Proceedings of the Royal Irish Academy, Section B: Biological, Geological, and Chemical Science*, 75 (1975), 275–84.

59. Bodleian, MS Eng. hist c. 11, fol. 59 (*EMLO* transcript).

60. Mitchell, pp. 280–1.

61. Mitchell, p. 281; popularly known as the Lingonberry or Cloudberry; see J. C. Ritchie, 'Vaccinium vitis-idaea L.', *Journal of Ecology*, 43/2 (1955), 701–8.

62. BL, Sloane MS 4063, fol. 48 (*EMLO* transcript).

63. E. C. Nelson, 'A History, mainly Nomenclatural, of St. Dabeoc's Heath', *Watsonia*, 23 (2000), 47–58 (pp. 47–9).

64. Campbell, p. 226.

65. Pearman, p. 84; BL, Sloane MS 4062, fols 286–7 (*EMLO* transcript).

66. Bodleian, MS Radcliffe Trust c. 1. fol. 54v (*EMLO* transcript).

67. MS Ashmole 1817a, fol. 451r (*EMLO* transcript).

68. Frank Emery, *Edward Lhuyd, F.R.S., 1660–1709* (Cardiff, 1971), p. 37.

69. Southampton Archives Services, D/M/1/2, pp. 135–6 (*EMLO* transcript).

70. For the geological importance of Ben Bulben, see Claire McAteer and Matthew Parkes, *The Geological Heritage of Sligo: An Audit of County Geological Sites in Sligo*, Geological Survey of Ireland (2004), *https://secure.dccae.gov.ie/GSI_DOWNLOAD/Geoheritage/Reports/Sligo_Audit.pdf*, accessed 18 March 2021.

71. For the Lhwyd–O'Flaherty correspondence, see Sharpe, *Letters*, pp. 203–306.

72. MS Ashmole 1814, fols 284–5 (*EMLO* transcript).

73. Considine, p. 133.

74. BL, Sloane MS 4063, fol. 48 (*EMLO* transcript).

75. O'Sullivan and O'Sullivan, p. 71, is the authoritative guide, together with *ELSH*, p. xiv.

76. R. L. Thomson, 'Edward Lhuyd in the Isle of Man?', in *Celtic Studies: Essays in Memory of Angus Matheson, 1912–1962*, ed. James Carney and David Greene (London, 1969), pp. 170–82 (pp. 178–9).

77. Considine, p. 142.

78. Dafydd Ifans and R. L. Thomson, 'Edward Lhuyd's *Geirieu Manaweg*', SC, 14/15 (1980), 129–67.

79. MS Ashmole 1817a, fol. 268v (*EMLO* transcript).

80. MS Ashmole 1814, fols 96–7 (*EMLO* transcript).

81. For Keigwin, apparently a native speaker of Cornish, see *ODNB*.

82. For a fuller discussion, see Brynley F. Roberts, 'Edward Lhwyd in Cornwall', SC, 53 (2019), 133–53.

83. See M. G. Smith, *'Fighting Joshua': A Study of the Career of Sir Jonathan Trelawny, Bart., 1650–1721, Bishop of Exeter, and Winchester* (Redruth, 1985), p. 155.

84. Raven, *Ray*, p. 257.

85. MS Ashmole 1816, fol. 373 (*EMLO* transcript).

86. MS Ashmole 1816, fol. 371 (*EMLO* transcript).

87. Printed in *PT*, 27 (1700), p. 527.

88. See *https://ukfossils.co.uk/category/cornwall/*, accessed 21 January 2021.

89. MS Ashmole 1815, fol. 344 (*EMLO* transcript); since the letter refers to the death of Sir John Aubrey on 15 September, Jones was probably beachcombing in late September or early October.

90. P. A. S. Pool, 'Cornish Drawings by Edward Lhuyd in the British Museum', *Cornish Archaeology*, 16 (1977), 139–42 (p. 140).

91. MS Ashmole 1817a, fols 116–17 (*EMLO* transcript).

92. Tonkin's note on Lhwyd's letter, printed in William Pryce, *Archæologia Cornu-Britannica* (Sherborne, 1790), p. [246]; 'Pennick' was John Pennecke, d. 1724, vicar of St Martin-juxta Looe in 1695, Chancellor of Exeter cathedral, 1706 (*Alum. Oxon.*).

93. Pryce, p. [247].

94. For the correspondence and detailed list of parishes, see Roberts, 'Cornwall', pp. 146–7.

95. See the map of parishes visited in Williams, p. 17.

96. Williams, pp. 15–16.

97. In the Cornish preface to the *Glossography*, Lhwyd mentions writing down some Cornish 'from the mouths of people in the West of Cornwall, in particular the parish of St Just', as well as Cornish words written out for him by four named 'Gentlemen', *Texts*, p. 153.

98. R. G. Maber, 'Celtic Prosody in Late Cornish: the *englyn* "An lavar kôth yu lavar guîr"', *BBCS*, 28/4 (1980), 593–8; I am particularly grateful to Dr Oliver Padel for discussing the Cornish *englyn*.

99. Daniel Huws, 'The Old Welsh *Englyn* in the Margam-Abbey Charter: A Lhuydian Joke', *Journal of Celtic Studies*, 4 (2004), 213–18, views it as a *jeu d'esprit* by one of Lhwyd's assistants.

100. P. T. J. Morgan, 'The Abbé Pezron and the Celts', *THSC* (1965), 286–95.

101. MS Ashmole 1814, fol. 90 (*EMLO* transcript).
102. Martin Lister, *A Journey to Paris in the Year 1698* (London, 1698), p. 96.
103. MS Ashmole 1817a, fol. 343 (*EMLO* transcript).
104. BL, Sloane MS 1472, fol. 48.
105. MS Ashmole 1815, fols 9–10 (*EMLO* transcript).
106. Royal Society, LBO/14, pp. 362–5 (*EMLO* transcript).
107. Daniel Le Bris, 'Les études linguistiques d'Edward Lhuyd en Bretagne en 1701', *La Bretagne linguistique*, 14 (2009), 175–93 (p. 176).
108. Pryce, p. [249].
109. Brynley F. Roberts, 'Edward Lhwyd's Collection of Printed Books', *The Bodleian Library Record*, 10 (1979), 112–27 (p. 125).
110. NLW, Peniarth MS 427, fols 74–5 (*EMLO* transcript).
111. For Lhwyd's Breton studies, see Rhisiart Hincks, 'Edward Llwyd a'r Llydaweg', *Y Naturiaethwr*, 11 (1984), 2–8; Rhisiart Hincks, *I Gadw Mamiaith mor Hen: Cyflwyniad i Ddechreuadau Ysgolheictod Llydaweg* (Llandysul, 1995).

Chapter 10

1. Ovenell, pp. 94–5.
2. Bodleian, MS Ashmole 1817b, fols 336–7 (*EMLO* transcript).
3. MS Ashmole 1816, fols 344–5 (*EMLO* transcript).
4. MS Ashmole 1815, fols 9–10 (*EMLO* transcript).
5. Ovenell, pp. 96–7 has Massey departing in November 1699; the correct date is November 1700, MS Ashmole 1816, fols 344–5 (*EMLO* transcript).
6. Ovenell, p. 93.
7. Roberts, 'Protégés', pp. 37–9.
8. MS Ashmole 1815, fol. 298 (*EMLO* transcript).
9. MS Ashmole 1817b, fol. 387 (*EMLO* transcript).
10. MS Ashmole 1815, fol. 298 (*EMLO* transcript).
11. Ovenell, p. 95.
12. MS Ashmole 1817b, fols 336–7 (*EMLO* transcript).
13. Bodleian, MS Register of Convocation, 1693–1703, fol. 235.
14. MS Ashmole 1817a, fols 345–6 (*EMLO* transcript).
15. Simcock, p. 37, n. 102.
16. MS Ashmole 1816, fols 117–18 (*EMLO* transcript).
17. Bodleian, MS Lister 3, fol. 157 (*EMLO* transcript).
18. MS Lister 36, fol. 128 (*EMLO* transcript).
19. MS Ashmole 1830, fol. 16 (*EMLO* transcript).
20. MS Ashmole 1816, fols 193–4 (*EMLO* transcript).
21. BL, Sloane MS 4062, fols 261, 262 (EMLO transcript).
22. MS Lister 3, fols 141–2 (*EMLO* transcript).
23. MS Ashmole 1816, fol. 197 (*EMLO* transcript).
24. MS Ashmole 1830, fols 7–8 (*EMLO* transcript).
25. MS Ashmole 1829, fols 155–6 (*EMLO* transcript).

26. MS Ashmole 1829, fols 111–12 (*EMLO* transcript).
27. MS Ashmole 1829, fols 141–2 (*EMLO* transcript).
28. *Texts*, pp. 49–51.
29. MS Ashmole 1814, fol. 60 (*EMLO* transcript).
30. *Texts*, p. 55.
31. *Texts*, p. 57.
32. BL, Sloane MS 4064, fol. 35 (*EMLO* transcript).
33. MS Ashmole 1817b, fol. 107 (*EMLO* transcript).
34. MS Ashmole 1817b, fol. 111 (*EMLO* transcript).
35. MS Ashmole 1817b, fol. 118 (*EMLO* transcript).
36. Roberts, 'Protégés', p. 44, n. 79.
37. MS Ashmole 1816, fols 300–1 (*EMLO* transcript).
38. Colin G. C. Tite, *The Manuscript Library of Sir Robert Cotton*, The Panizzi Lectures, 1993 (London, 1994), p. 37; for the layout and appearance of the library, see pp. 79–100.
39. MS Ashmole 1815, fol. 303 (*EMLO* transcript).
40. For the *Vocabularium*, see Alderik H. Blom, 'The Welsh Glosses in the Vocabularium Cornicum', *Cambrian Medieval Celtic Studies*, 57 (2009), 23–40; Barry Lewis, 'A Possible Provenance for the Old Cornish Vocabulary?', *Cambrian Medieval Celtic Studies*, 73 (2017), 1–14; Oliver Padel, 'The Nature and Date of the Old Cornish Vocabulary', *ZCP*, 61 (2014), 173–99.
41. BL, Stowe MS 749, fol. 4 (*EMLO* transcript).
42. *Texts*, pp. 154–5.
43. W. Pryce, *Archæologia Cornu-Britannica* (Sherborne, 1790), p. 253.
44. MS Ashmole 1815, fol. 335 (*EMLO* transcript).
45. MS Ashmole 1815, fol. 318 (*EMLO* transcript).
46. MS Ashmole 1815, fol. 331 (*EMLO* transcript).
47. Roberts, 'Protégés', p. 47.
48. BL, Sloane MS 4063, fols 185–6 (*EMLO* transcript).
49. Pryce, p. 253. According to the *Clergy of the Church of England Database*, Jones became curate of Clun with Chapel Lawn and Mainstone on 18 September 1702. A draft letter in Gilli's hand from December 1702 says Jones was 'preferrd near Bishops Castle to a place of Forty Pound a year but dy'd of his Rupture almost as soon as he came to it', Roberts, 'Protégés', p. 47, n. 87.
50. MS Ashmole 1815, fol.142 (*EMLO* transcript).
51. Roberts, 'Cymraeg', pp. 221–2.
52. Ovenell, p. 101.
53. MS Ashmole 1817b, fol. 207 (*EMLO* transcript).
54. MS Ashmole 1817a, fol. 186 (*EMLO* transcript).
55. MS Ashmole 1817a, fols 187–8 (*EMLO* transcript).
56. For Lhwyd's Cardiganshire relatives, see Roberts, 'Ceredigion', p. 62 and n. 16.
57. MS Ashmole 1817a, fols 189–90 (*EMLO* transcript).
58. MS Ashmole 1817a, fols 191–2 (*EMLO* transcript).
59. Roberts, 'Edward Lhwyd', 64.

60. For Sage Lloyd, see Roberts, 'Carmarthenshire', pp. 34–5.

61. Edward Bernard's *Catalogi librorum manuscriptorum Angliae et Hiberniae* (1697); see J. C. T. Oates, *Cambridge University Library: A History. From the Beginnings to the Copyright Act of Queen Anne* (Cambridge, 1986), pp. 484–5.

62. Bodleian, MS Radcliffe Trust c. 1, fol. 94.

63. On the manuscript, see Huws, *Repertory*, I; for the poems, see Ifor Williams, 'The Juvencus Poems', in *The Beginnings of Welsh Poetry: Studies*, ed. Rachel Bromwich (Cardiff, 1972), pp. 89–121.

64. Brynley F. Roberts., 'Translating Old Welsh: The First Attempts', *ZCP*, 49–50 (1997), 760–77 (p. 776).

65. Hearne, X, 288.

66. J. C. T. Oates, 'Notes on the Later History of the Oldest Manuscript of Welsh Poetry: The Cambridge Juvencus', *Cambridge Medieval Celtic Studies*, 3 (1982), 81–7.

67. *Glossography*, p. 221.

68. For Baxter see *DWB*.

69. NLW, MS 309E, pp. 43–5 (*EMLO* transcript).

70. Roberts, 'Translating', pp. 775–6.

71. Brynley F. Roberts, 'The Discovery of Old Welsh', *Historiographia Linguistica*, 26 (1999), 1–21.

72. MS Ashmole 1817b. fols 197–8 (*EMLO* transcript).

73. MS Ashmole 1817b, fols 199–200 (*EMLO* transcript).

74. Richard Ellis, 'Llythyrau Llafurwr', *Cymru*, 25 (1903), 189–93 (p. 189).

75. MS Ashmole 1815, fols 184–5 (*EMLO* transcript).

76. *Texts*, pp. 8–9.

77. Lhwyd to Humphrey Humphreys, 13 October 1706, in E. Gwynne Jones, 'The Family Papers of Owen and Stanley of Penrhos, Holyhead', *BBCS*, 18/2 (1957), 99–115 (p. 114).

78. MS Radcliffe Trust c. 2, fols 39–40 (*EMLO* transcript).

79. MS Ashmole 1815, fol. 47 (*EMLO* transcript).

80. *PT*, 25 (1706–7), 2438–44.

81. BL, Sloane MS 4067, fols 26–7 (*EMLO* transcript).

82. MS Ashmole 1817a, fols 400–1 (*EMLO* transcript).

83. NLW, Peniarth MS 427, fol. 80 (*EMLO* transcript).

84. MS Ashmole 1815, fols 188–9 (*EMLO* transcript).

85. NLW, Llanstephan MS 84, is available online at *http://hdl.handle. net/10107/4398124*.

86. David Cram, 'Edward Lhuyd's *Archæologia Britannica*: Method and Madness in Early Modern Comparative Philology', *WHR*, 25/1 (2010), 75–96 (pp. 84–5).

87. *Texts*, pp. 14–17.

88. *Texts*, pp. 117–27.

89. See poems by Colin Campbell, *Texts*, pp. 74–7, and John Maclean, *Texts*, pp. 94–7.

90. *Texts*, pp. 121–5.

91. G, p. 511, to Humphrey Foulkes, May [1707]; cf. G, p. 482, letter to Henry Rowlands, [February 1703]; cf. Colin Campbell's poem and following notes.

92. For Pezron and his influence, see Caryl Davies, *Adfeilion Babel: Agweddau ar Syniadaeth Ieithyddol y Ddeunawfed Ganrif* (Cardiff, 2000), pp. 60–92; Michael A. Morse, *How the Celts Came to Britain: Druids, Ancient Skulls and the Birth of Archaeology* (Stroud, 2005), pp. 22–31.

93. MS Ashmole 1816, fols 142–3 (*EMLO* transcript).

94. Considine, p. 145.

95. See Considine, chapters 14–16, for the development of Lhwyd's ideas.

96. For Lhwyd's thinking about a 'Celtic family' of languages, see Considine, p. 144.

97. Brynley F. Roberts, 'Edward Lhuyd and Celtic Linguistics', in *Proceedings of the Seventh International Congress of Celtic Studies*, ed. D. Ellis Evans, John G. Griffith and E. M. Jope (Oxford, 1986), pp. 1–9 (p. 8).

98. MS Ashmole 1815, fol. 152 (*EMLO* transcript corrected from image).

99. Probably Henry Aldrich (1648–1710), Dean of Christ Church, Oxford; the book never appeared.

100. MS Radcliffe Trust c. 2, fol. 51 (*EMLO* transcript).

101. Johann Jakob Scheuchzer, ΟΥΡΕΣΙΦΟΙΤΕΣ *sive itinera alpina tria* (London, 1708).

102. MS Ashmole 1817b, fols 153–4 (*EMLO* transcript).

103. MS Ashmole 1817a, fols 488–9 (*EMLO* transcript).

104. MS Ashmole 1817b, fol. 155 (*EMLO* transcript).

105. MS Ashmole 1817a, fols 362–3 (*EMLO* transcript).

106. For a picture and brief description, see *https://www.hikr.org/gallery/photo3358260. html?piz_id=8150*, accessed 15 January 2021.

107. *PT*, 26 (1708), 143–67 (pp. 157–8).

108. MS Ashmole 1817b, fols 157–8 (*EMLO* transcript).

109. See Joseph M. Levine, *Dr Woodward's Shield: History, Science, and Satire in Augustan England* (Ithaca and London, 1977), pp. 89–91.

110. MS Ashmole 1817b, fols 159–60 (*EMLO* transcript).

111. *The London Diaries of William Nicolson, Bishop of Carlisle, 1702–1718*, ed. Clyve Jones and Geoffrey Holmes (Oxford, 1985), p. 493.

112. Hearne, II, 201.

113. *Alum. Oxon.*

114. Hearne, II, 172.

115. Bodleian, MSS Top. Oxon. e. 281, is the Club's minute book 1694–1775.

116. Published in Alexander Chalmers, *General Biographical Dictionary*, 20 (London, 1815), 232–6. See Roberts, 'Memoirs', pp. 85–6.

117. Roberts, 'Barddoniaeth', 40–2; Lhwyd may have attempted an *englyn* earlier than this; see p. 40.

118. See *ODNB*, which discusses the popularity of *Muscipula*.

119. Brynley F. Roberts, 'Dialedd Taffy (*Hoglandiae Descriptio* Thomas Richards)', *Ysgrifau Beirniadol*, 25 (1999), 61–78.

120. Roberts, 'Llythyrau', p. 197.

121. MS Ashmole 1817b, fol. 387 (*EMLO* transcript).

122. MS Ashmole 1816, fols 142–3 (*EMLO* transcripts).

123. MS Ashmole 1817b, fol. 118 (*EMLO* transcript).

124. Hearne, II, 219.
125. Contemporaries blamed Lhwyd's sleeping in a damp room in the Museum (see Chapter 3). Dr Brian H. Davies suggests in 'Edward Lhuyd and Asbestos Paper', *The Quarterly*, 47 (2003), 10–11, that Lhwyd's death may have resulted from his exposure to asbestos. Mesothelioma is a condition characterised by a dry cough, respiratory problems and fatigue, typically manifesting itself some twenty-five to forty years after exposure; Lhwyd's asbestos paper experiments were made *c*.1684; he died about twenty-five years later.
126. Hearne, II, 218–19, n. 1.
127. Huddesford, pp. 142, 172.
128. Hearne, IV, 44, 48.
129. Brynley F. Roberts, 'Pedwar Portread o Edward Lhuyd', *NLWJ*, 21/1 (1979), 40–2.

Chapter 11

1. For a more detailed account of several matters discussed in this chapter, see Brynley F. Roberts, 'Etifeddion Edward Lhuyd', *WHR*, 25/1 (2010), 97–119.
2. Uffenbach, pp. 30–1.
3. Hearne, II, 221.
4. *Texts*, pp. 25–6.
5. Hearne, II, 225.
6. Brynley F. Roberts, 'Edward Lhuyd's Debts', *BBCS*, 26/3 (1975), 353–9.
7. Roberts, 'Books', pp. 119–21.
8. Brynley F. Roberts, 'A Note on the Ashmolean Collection of Letters addressed to Edward Lhuyd', *WHR*, 7/2 (1975), 179–85; for Huddesford, see *ODNB*.
9. See Eiluned Rees and Gwyn Walters, 'The Dispersion of the Manuscripts of Edward Lhuyd', *WHR*, 7/2 (1974), 148–78; Huws, *Repertory*, s.n. Edward Lhuyd, contains the fullest account of the Lhwyd archive.
10. These are included in *Camden's Wales*.
11. *Texts*, pp. 216–31.
12. Walter Moyle, *The Works of Walter Moyle* … 2 vols (London, 1726), I, 183–4, 195–6, 201–2, 236–41, 248–51.
13. They are more conveniently grouped together in the posthumous second edition of 1766.
14. Mary Ellis, 'Angharad Llwyd 1780–1866', *Flintshire Historical Society Journal*, 26 (1973–4), 52–95 (pp. 79–80); see Huws, *Repertory*, s.n. John Lloyd of Caerwys.
15. See for example the *horti sicci* noted in *The Sloane Herbarium: An Annotated List of the* Horti Sicci *Composing it: With Biographical Accounts of the Principal Contributors* (London, 1958), pp. 155–7.
16. Pearman, pp. 65–6; these include *Cystopteris alpina*, recently attributed to Lhwyd, p. 170.
17. John Ray, *Synopsis Methodica Stirpium Britannicarum* … 3rd edn (London, 1724), facsimile with an introduction by William T. Stearn (London, 1973), p. 32.

18. Arthur MacGregor and A. Turner, 'The Ashmolean Museum', in *The History of the University of Oxford*, V: *The Eighteenth Century*, ed. L. S. Sutherland and L. G. Mitchell (Oxford, 1986), pp. 639–58 (p. 647).

19. Uffenbach, p. 49.

20. Melvin E. Jahn, 'The Old Ashmolean Museum and the Lhwyd Collections', *JSBNS*, 4 (1966), 244–8 (p. 245).

21. MacGregor and Turner, p. 653, n. 4.

22. Gunther, *Biological*, pp. 374–5.

23. Described and illustrated on p. 2 of *Edward Lhwyd: learning more* at *https://www.oum.ox.ac.uk/learning/pdfs/lhwyd.pdf*.

24. Bodleian, MS Carte, 269, fol. 135.

25. Roy Porter, *The Making of Geology: Earth Science in Britain 1660–1815* (Cambridge, 1977), p. 92.

26. Consuelo Sendino, 'The Hans Sloane Fossil Collection at the Natural History Museum, London', *Deposits Mag*, 47 (2016), 13–17.

27. BL, Sloane MS 4039, fols 255–6 (*EMLO* transcript).

28. G, pp. 558–60, followed by plates illustrating the two survivors and all the specimens in the Oriel collection.

29. A fellow of Oriel and enthusiastic botanist. Lhwyd thought he could have seen the *Ichnographia* through the press had it been printed in Oxford, Ovenell, p. 86.

30. For an illustrated account see *https://hsm.ox.ac.uk/collections-online#/item/hsm-catalogue-9444*, accessed 21 January 2121.

31. See Anna Marie Roos and Edwin Rose, 'Lives and Afterlives of the *Lithophylacii Britannici ichnographia* (1699), the First Illustrated Field Guide to English Fossils', *Nuncius*, 23 (2018), 505–36; the edition size given there of 120 (p. 534) is an error, apparently based on the information for the first edition being reproduced in the second; Ovenell (p. 149) correctly gives 300 as the edition size.

32. Roos and Rose, p. 532.

33. Roberts, 'Memoirs'.

34. The complete text was published in *AC*, 1846–9.

35. See John Davies, *Bywyd a Gwaith Moses Williams (1685–1743)* (Cardiff, 1937), p. 21.

36. NLW, MS 9628E, item 42; see Mary Burdett-Jones, 'Building the Palace: Dr Humphrey Foulkes's Attempt to Continue Edward Lhuyd's Work', *Denbighshire Historical Society Transactions*, 58 (2010), 11–22.

37. For Welsh lexicography, see G. Angharad Fychan, Andrew Hawke and Ann Parry Owen (eds), *Trysordy'r Iaith* (forthcoming).

38. Porter, pp. 91–4.

39. A recent discussion which surveys earlier studies is Ronald Hutton, *Blood and Mistletoe: The History of the Druids in Britain* (New Haven and London, 2009).

40. Piggott, pp. 116–18.

41. Sam Smiles, *The Image of Antiquity: Ancient Britain and the Romantic Imagination* (New Haven and London, 1994), Chapter 5.

42. Smiles, p. 55.

43. *Texts*, pp. 84–7.

44. Garfield H. Hughes, 'Iaco ab Dewi: Rhai Ystyriaethau', *NLWJ*, 3/1–2 (1943), 51–5 (p. 52); the *englynion* are printed in Garfield H. Hughes, *Iaco ab Dewi (1648–1722)* (Cardiff, 1953), p. 33.

45. Garfield H. Hughes, 'Dafydd Manuel', *Llên Cymru*, 6 (1960), 26–34; *Texts*, pp. 242–3.

46. *Texts*, pp. 244–52.

47. R. T. Jenkins and Helen Ramage, *A History of the Honourable Society of Cymmrodorion and of the Gwyneddigion and Cymreigyddion Societies*, *Y Cymmrodor*, 50 (1951), 241–4.

48. The first part was published in 1878.

49. Hugh Owen (ed.), *Additional Letters of the Morrises of Anglesey (1735–1786)*, *Y Cymmrodor*, 49/1 (1947), I, 206.

50. Letter from Gwallter Mechain to Iolo Morganwg, 15 March 1793: *The Correspondence of Iolo Morganwg*, I: *1770–96*, ed. Geraint H. Jenkins, Ffion Mair Jones and David Ceri Jones (Cardiff, 2007), pp. 554–7, no. 248.

51. Some of these copies of Lhwyd's correspondence are in NLW 1663C.

52. Possibly Bodleian, MS Rawl. B 464.

53. *Correspondence of Iolo Morganwg*, I, 555.

54. Reprinted in *The English Works of the Rev. Walter Davies, M.A. (Gwallter Mechain)*, ed. D. Silvan Evans (Carmarthen and London, 1868), pp. 1–134.

55. See Brynley F. Roberts, 'Scholarly publishing 1820–1922', in *A Nation and its Books: A History of the Book in Wales*, ed. Philip Henry Jones and Eiluned Rees (Aberystwyth, 1998), pp. 221–35.

56. Tourneur, *Esquisse*, p. 207.

57. G. W. Leibniz, *Collectanea etymologica, illvstrationi lingvarvm, veteris Celticæ ... inservientia* (Hanover, 1717), p. 153; his reference to a newly published work in England on 'antiqua lingua Britannica', p. 147, may be to the *Archæologia*.

58. See Francis Shaw, 'The Background to *Grammatica Celtica*', *Celtica*, 3 (1956), 1–16 (p. 7); see also Davies, *Adfeilion*, p. 68 and Chapters 2 and 3.

59. Davies, Chapter 8.

60. Davies, Chapter 10.

61. I am grateful to Dr Löffler for a copy of the valuable paper which she gave in the conference on Edward Lhwyd in Aberystwyth in 2009. Stephens's library contained *Archæologia Britannica* and Nicholas Owens's *British Remains*; Stephens referred to Lhwyd's work in, e.g., NLW, MSS 934B, fols 55, 65, 76; 907C, fols 17, 112.

62. *Bye-Gones* (1896), 363.

63. John Morris-Jones, 'Edward Lhwyd', *Y Traethodydd*, 49 (1893), 465–75; John Morris-Jones, *Edward Lhwyd: Inaugural Address Delivered at the University College of North Wales, Session 1893–94* (Caernarfon, 1894). He had already shown the significance of Lhwyd for comparative linguistics in his article 'Cymraeg', *Y Gwyddoniadur Cymreig*, 2nd edn (Denbigh, 1892), III, 48–79 (p. 74). The entry on Edward Lhwyd in *Y Gwyddoniadur Cymreig*, 2nd edn, VII, 52–3, pays scant attention to his linguistic work.

64. *THSC*, 1906–7 (1908), 1–51.

65. Roberts, 'Richard Ellis, M.A.', pp. 153–4.

66. Dyfed Ellis Gruffudd, 'Nid oes dim newydd dan yr Haul: Coffâu Edward Lhwyd', *Y Naturiaethwr*, 2/3 (1998), 12–13.

67. *https://cymdeithasedwardllwyd.cymru/hanes-y-gymdeithas*.

68. Glyn Daniel, 'Edward Lhwyd, Antiquary and Archaeologist', *WHR*, 3/4 (1967), 345–59 (p. 351).

69. Nancy Edwards, 'Edward Lhuyd and the Origins of Early Medieval Celtic Archaeology', *Antiquaries Journal*, 87 (2007), 165–96.

70. Jeremy Knight, 'Welsh Stones and Oxford Scholars: Three Rediscoveries', in *The Afterlife of Inscriptions*, ed. Alison Cooley (London, 2000), pp. 91–101; Nancy Edwards, 'Rethinking the Pillar of Eliseg', *Antiquaries Journal*, 89 (2009), 143–77.

71. Parry, Chapter 12.

72. James January-McCann, 'Bywyd newydd i'r *Parochialia*', *Y Naturiaethwr*, 3/14 (2020), 8–10.

73. *PT*, 48 (1753–4), 286–7.

74. Christopher Duffin, 'The Earliest Published Records of Coprolites', *New Mexico Museum of Natural History and Science Bulletin*, 57 (2012), 25–8 (p. 26).

75. Christophe Hendrickx, Scott A. Hartman and Octavio Mateus, 'An Overview of Non-Avian Theropod Discoveries and Classification', *PalArch's Journal of Vertebrate Palaeontology*, 12/1 (2015), 1–73 (p. 2).

76. Hearne, I, 244.

77. John Leland, *The Itinerary of John Leland … *ed. Thomas Hearne, 9 vols (Oxford, 1710–12), II, iii.

BIBLIOGRAPHY

Adams, Frank Dawson, *The Birth and Development of the Geological Sciences* (New York, 1938).

Allan, Mea, *The Tradescants, their Plants, Gardens and Museum, 1570–1662* (London, 1964).

Allen, David Elliston, *The Naturalist in Britain: A Social History*, 2nd edn (Princeton, 1994).

Andrews, Henry N., *The Fossil Hunters: In Search of Ancient Plants* (Ithaca and London, 1980).

Arbuthnot, John, *An Examination of Dr. Woodward's Account of the Deluge, &c. with a Comparison between STENO'S Philosophy and the DOCTOR'S, in the Case of Marine Bodies dug out of the Earth. By J. A. M.D. With a LETTER to the Author concerning An ABSTRACT of AGOSTINO SCILLA'S Book on the same Subject ... By W. W. F.R.S.* (London, 1697).

Arnold, Ken, *Cabinets for the Curious: Looking Back at Early English Museums* (Aldershot, 2006).

Aubrey, John, *Brief Lives, with An Apparatus for the Lives of our English Mathematical Writers*, ed. Kate Bennett, 2 vols (Oxford, 2015).

Barry, Jonathan, *Witchcraft and Demonology in South-West England, 1640–1789* (London, 2012).

Bartos, Jim, 'The Spirituall Orchard: God, Garden and Landscape in Seventeenth-Century England before the Restoration', *Garden History*, 38/2 (2010), 177–93.

Bassett, Michael G., *'Formed Stones', Folklore and Fossils*, Geological Series No. 1 (Cardiff, 1982).

Bassett, Michael G. and Diane Edwards, *Fossil Plants from Wales*, Geological Series No. 2 (Cardiff, 1982).

Batey, Mavis, *Oxford Gardens: The University's Influence on Garden History* (Amersham, 1982).

Bennet, J. A., S. A Johnston and A. V. Simcock, *Solomon's House in Oxford: New Finds from the First Museum* (Oxford, 2000).

Blom, Alderik H., 'The Welsh Glosses in the Vocabularium Cornicum', *Cambrian Medieval Celtic Studies*, 57 (2009), 23–40.

Boon, George C., *Cardiganshire Silver and the Aberystwyth Mint in Peace and War* (Cardiff, 1981).

Boon, George C., 'The Llanymynech Roman Imperial Treasure Trove', *The Numismatic Chronicle*, 7th series, 6 (1966), 155–6.

Bowen, D. J., 'Croesoswallt y Beirdd', *Y Traethodydd*, 135 (1980), 137–43.

Bowen, D. J., 'Cynefin Wiliam Llŷn', *Barn* (Gorff./Awst 1980), 206–8.

Bracher, Terry and Roger Emmett, *Shropshire in the Civil War* ([Shrewsbury], 2000).

Briggs, C. Stephen, 'Edward Lhuyd's Expeditions through Ulster, 1699 and 1700', in *The Modern Traveller to Our Past: Festschrift in Honour of Ann Hamlin*, ed. Marion Meek (n.p., 2006), pp. 384–93.

Burdett-Jones, Mary, 'Building the Palace: Dr Humphrey Foulkes's Attempt to Continue Edward Lhuyd's Work', *Denbighshire Historical Society Transactions*, 58 (2010), 11–22.

Burnby, J., 'Some Early London Physic Gardens', *Pharmaceutical Historian*, 24/4 (1994), 2–8.

Camden's Wales, Being the Welsh Chapters Taken from Edmund Gibson's Revised & Enlarged Edition of William Camden's BRITANNIA (1722) Translated from the Latin, with Additions, by Edward Lhuyd, with Maps executed by Robert Morden, with an introductory essay by Gwyn Walters (Carmarthen, 1984).

Campbell, J. L. (ed.), *A Collection of Highland Rites and Customes copied by Edward Lhuyd from the Manuscript of the Rev. James Kirkwood ... and annotated by him with the aid of the Rev. John Beaton* (Cambridge, 1975).

Campbell, J. L., 'The Contribution of Edward Lhuyd to the Study of Scottish Gaelic', *THSC* (1962), 77–80.

Campbell, J. L., 'The Tour of Edward Lhuyd in Ireland in 1699 and 1700', *Celtica*, 5 (1960), 218–28.

Campbell, J. L., 'Unpublished Letters by Edward Lhuyd in the National Library of Scotland', *Celtica*, 11 (1976), 34–42.

Campbell, J. L. and Derick Thomson, *Edward Lhuyd in the Scottish Highlands 1699–1700* (Oxford, 1963).

Cathrall, William, *The History of Oswestry ...* (Oswestry, [1855]).

Chapman, Allan, 'From Alchemy to Airpumps: The Foundations of Oxford Chemistry to 1700', in *Chemistry at Oxford: A History from 1600 to 2005*, ed. Robert J. P. Williams, John S. Rowlinson and Allan Chapman (Cambridge, 2009), pp. 17–51.

Chater, A. O., 'An Unpublished Botanical Notebook of Edward Llwyd', *Botanical Society of the British Isles Welsh Bulletin*, 40 (1984), 4–15.

Chater, A. O., 'Lloydia serotina', *Welsh Bulletin of the Botanical Society of the British Isles*, 38 (July 1983), 3–7.

Cleal, Christopher J. and Barry A. Thomas, *Plant Fossils of the British Coal Measures*, Palaeontological Association Field Guides to Fossils, 6 (London, 1994).

Cliffe, J. T., 'The Cromwellian Decimation Tax of 1655, the Assessment Lists', in *Seventeenth-Century Political and Financial Papers*, Camden Miscellany 23, Camden 5th Series, vol. 7 (Cambridge, 1996), pp. 403–92.

Considine, John, *Small Dictionaries and Curiosity: Lexicography and Fieldwork in Post-medieval Europe* (Oxford, 2017).

Cook, Alan, *Edmond Halley: Charting the Heavens and the Seas* (Oxford, 1998).

Cram, David, 'Edward Lhuyd's *Archæologia Britannica*: Method and Madness in Early Modern Comparative Philology', *WHR*, 25/1 (2010), 75–96.

Cramsie, John, *British Travellers and the Encounter with Britain, 1450–1700* (Woodbridge, 2015).

Cutler, Alan, *The Seashell on the Mountaintop* (New York, 2003).

Daniel, Glyn, 'Edward Lhuyd, Antiquary and Archaeologist', *WHR*, 3/4 (1967), 345–59.

Davies, Brian H., 'Edward Lhuyd and Asbestos Paper', *The Quarterly*, 47 (2003), 10–11.

Davies, Caryl, *Adfeilion Babel: Agweddau ar Syniadaeth Ieithyddol y Ddeunawfed Ganrif* (Cardiff, 2000).

Davies, Caryl and Mary Burdett-Jones, 'Cyfraniad Humphrey Foulkes at *Archæologia Britannica* Edward Lhuyd', *Y Llyfr yng Nghymru*, 8 (2007), 7–32.

Davies, Gordon L., 'Early British Geomorphology 1578–1705', *Geographical Journal*, 132/2 (1966), 252–62.

Davies, John, *Antiquæ Linguæ Britannicæ ... Dictionarium Duplex* (London, 1632).

Davies, John, *Bywyd a Gwaith Moses Williams (1685–1743)* (Cardiff, 1937).

Davies, Walter, *The English Works of the Rev. Walter Davies, M.A. (Gwallter Mechain)*, ed. D. Silvan Evans (Carmarthen and London, 1868).

Deming, David, 'Edmond Halley's Contributions to Hydrogeology', *Groundwater*, 59/1 (2021), 146–52.

Dodd, A. H., *Studies in Stuart Wales* (Cardiff, 1952).

Dodd, A. H., 'The Early Life of Edward Lhuyd', *NLWJ*, 6/3 (1950), 305–6.

Duffin, Christopher, 'The Earliest Published Records of Coprolites', *New Mexico Museum of Natural History and Science Bulletin*, 57 (2012), 25–8.

Duffin, Christopher and Jane P. Davidson, 'Geology and the Dark Side', *Proceedings of the Geologists' Association*, 122 (2011), 7–15.

Early Modern Letters Online, Cultures of Knowledge, *http://emlo.bodleian.ox.ac.uk*.

Edwards, Ifan ap Owen, *A Catalogue of Star Chamber Proceedings relating to Wales*, Board of Celtic Studies, University of Wales History and Law Series, No. I (Cardiff, 1929).

Edwards, Nancy, *A Corpus of Early Medieval Inscribed Stones and Stone Sculpture in Wales*, vol. 2: *South-West Wales* (Cardiff, 2007).

Edwards, Nancy, *A Corpus of Early Medieval Inscribed Stones and Stone Sculpture in Wales*, vol. 3: *North Wales* (Cardiff, 2013).

Edwards, Nancy, 'Edward Lhwyd: An Archaeologist's View', *WHR*, 25/1 (2010), 20–50.

Edwards, Nancy, 'Edward Lhuyd and the Origins of Early Medieval Celtic Archaeology', *Antiquaries Journal*, 87 (2007), 165–96.

Edwards, Nancy, 'Rethinking the Pillar of Eliseg', *Antiquaries Journal*, 89 (2009), 143–77.

Ellis, Mary, 'Angharad Llwyd 1780–1866', *Flintshire Historical Society Journal*, 26 (1973–4), 52–95; 27 (1975–6), 43–84.

Ellis, Richard, 'Llythyrau Llafurwr', *Cymru*, 25 (1903), 189–93.

Ellis, Richard, 'Some incidents in the life of Edward Lhuyd', *THSC*, (1906–7), 1–51.

Emery, F. V., 'A Map of Edward Lhuyd's *Parochial Queries in Order to a Geographical Dictionary, &c., of Wales* (1696)', *THSC* (1958), 41–53.

Emery, Frank, 'A New Account of Snowdonia, 1693, Written for Edward Lhuyd', *NLWJ*, 18/4 (1974), 405–17.

Emery, F. V., *Edward Lhuyd, F.R.S. (1660–1709)* (Cardiff, 1971).

Emery, F. V., 'Edward Lhuyd and Snowdonia', *Nature in Wales*, NS, 4/1–2 (1986), 3–11.

Emery, F.V., 'Edward Lhuyd and some of his Glamorgan Correspondents: A View of Gower in the 1690s', *THSC* (1965), 59–114.

Emery, F. V., 'Edward Lhuyd and the 1695 *Britannia*', *Antiquity*, 32 (1958), 179–82.

Evans, Dewi W. and Brynley F. Roberts, *Edward Lhwyd, 1660–1709: Llyfryddiaeth a Chyfarwyddiadur, A Bibliography and Readers' Guide* (Aberystwyth, 2009).

Evans, R. E., *Llanfihangel Genau'r Glyn: The History of a Community* (Llandre, 2010).

Fairer, David. 'Anglo-Saxon Studies', in *The History of the University of Oxford*, vol. V: *The Eighteenth Century*, ed. L. S. Sutherland and L. G. Mitchell (Oxford, 1986), pp. 807–29.

Feingold, Mordechai, 'The Mathematical Sciences and New Philosophies', in *The History of the University of Oxford*, vol. IV: *Seventeenth-Century Oxford*, ed. Nicholas Tyacke (Oxford, 1997), 359–448.

Fenton, Richard, *Tours in Wales (1804–1813)*, ed. John Fisher (London, 1917).

Feola, Vittoria and Scott Mandelbrote, 'The Learned Press: Geography, Science, and Mathematics', in *The History of Oxford University Press*, vol. I: *Beginnings to 1780*, ed. Ian Gadd (Oxford, 2013), pp. 317–57.

Fisher, G. W., *Annals of Shrewsbury School*, rev. J. Spencer Hill (London, 1899).

Floud, Roderick, *An Economic History of the English Garden* (n.p., 2019).

Fox, Adam, 'Printed questionnaires, and research networks, 1650–1800', *Historical Journal*, 53 (2010), 593–621.

Gibson, William, 'The correspondence of Edward Lhuyd and John Wynne', *Flintshire Historical Society Journal*, 32 (1989), 16–24.

Goodrum, Matthew R., 'Atomism, Atheism, and the Spontaneous Generation of Human Beings: The Debate over a Natural Origin of the First Humans in Seventeenth-Century Britain', *Journal of the History of Ideas*, 63/2 (2002), 207–24.

Griffith, John Edward, *Pedigrees of Anglesey and Carnarvonshire Families with their Collateral Branches in Denbighshire, Merionethshire and other parts* (Horncastle, 1914).

Gruffudd, Dyfed Ellis, 'Nid oes dim newydd dan yr Haul: Coffáu Edward Lhwyd', *Y Naturiaethwr*, 2/3 (1998), 12–13.

Gunther, R. T., *Early British Botanists and their Gardens, based on Unpublished Writings of Goodyer, Tradescant, and others* (Oxford, 1922).

Gunther, R. T., *Early Science in Oxford*, vol. III, part I: *The Biological Sciences*; part II: *The Biological Collections* (Oxford, 1925).

Gunther, R.T., *Early Science in Oxford*, vol. IV: *The Philosophical Society* (Oxford, 1925).

Gunther, R. T., *Early Science in Oxford*, vol. XII: *Dr Plot and the Correspondence of the Philosophical Society of Oxford* (Oxford, 1939).

Gunther, R. T., *Early Science in Oxford*, vol. XIV: *Life and Letters of Edward Lhwyd* (Oxford, 1945).

Gwaith Lewys Glyn Cothi, ed. Dafydd Johnston (Cardiff, 1995).

Gwaith Tudur Aled, ed. T. Gwynn Jones, 2 vols (Cardiff, 1926).

Haber, Francis C., *The Age of the World: Moses to Darwin* (Baltimore, 1959).

Harris, John, *Remarks on some Late Papers relating to the Universal Deluge …* (London, 1697).

Harris, Stephen A., *Oxford Botanic Garden & Arboretum: A Brief History* (Oxford, 2017).

Harvey, John, *Early Gardening Catalogues* (London and Chichester, 1972).

Harvey, John, *Early Nurserymen* (Chichester, 1974).

Haslam, Richard, 'Bodysgallen: A Renaissance Garden Survival?', *Garden History*, 34/1 (2006), 132–44.

Hayman, Richard and Wendy Horton, *Manod Bach – Y Garnedd: An Archaeological Survey* (n.p., 2014).

[Hearne, Thomas], *Remarks and Collections of Thomas Hearne*, ed. C. E. Doble, D. W. Rannie and H. E. Salter, 11 vols (Oxford, 1885–1921).

Hellyer, Marcus, 'The Pocket Museum: Edward Lhwyd's *Lithophylacium*', *ANH*, 23 (1966), 43–60.

Hendrickx, Christophe, Scott A. Hartman and Octavio Mateus, 'An Overview of Non-Avian Theropod Discoveries and Classification', *PalArch's Journal of Vertebrate Palaeontology*, 12/1 (2015), 1–73.

Hincks, Rhisiart, 'Edward Llwyd a'r Llydaweg', *Y Naturiaethwr*, 11 (1984).

Hincks, Rhisiart, *I Gadw Mamiaith mor Hen: Cyflwyniad i Ddechreuadau Ysgolheictod Llydaweg* (Llandysul, 1995).

Hoppen, K. Theodore, *The Common Scientist in the Seventeenth Century: A Study of the Dublin Philosophical Society 1683–1708* (London, 1970).

Houghton, Walter E. Jr, 'The English Virtuoso in the Seventeenth Century', *Journal of the History of Ideas*, 3/1–2 (1942), 51–73, 190–219.

[Huddesford, William], 'Memoirs of the life of Edward Lhwyd, M.A.', in Nicholas Owen, *British Remains, or a Collection of Antiquities Relating to the Britons* (London, 1777), pp. 131–84.

Hughes, Garfield H., 'Dafydd Manuel', *Llên Cymru*, 6 (1960), 26–34.

Hughes, Garfield H., *Iaco ab Dewi (1648–1722)* (Cardiff, 1953).

Hughes, Garfield H., 'Iaco ab Dewi: Rhai Ystyriaethau', *NLWJ*, 3/1–2 (1943), 51–5.

Hughes, R. Elwyn, 'Blodeuyn Edward Llwyd', *Y Naturiaethwr*, 13 (1985), 16–18.

Hunter, Michael, *Elias Ashmole 1617–1692: The Founder of the Ashmolean Museum and his World* (Oxford, 1983).

Hunter, Michael, *John Aubrey and the Realm of Learning* (London, 1975).

Hunter, Michael, *Science and Society in Restoration England* (Cambridge, 1981).

Hunter, Michael, *The Decline of Magic: Britain in the Enlightenment* (New Haven and London, 2020).

Hutton, Ronald, *Blood and Mistletoe: The History of the Druids in Britain* (New Haven and London, 2009).

Huws, Daniel, *A Repertory of Welsh Manuscripts and Scribes, c.800–c.1800*, 3 vols (Aberystwyth, 2022).

Huws, Daniel, 'The Old Welsh *Englyn* in the Margam-Abbey Charter: A Lhuydian Joke', *Journal of Celtic Studies*, 4 (2004), 213–18.

Huws, Daniel, 'Wiliam Llŷn, Rhys Cain a Stryd Wylw', *NLWJ*, 18/1 (1973), 147–8.

Ifans, Dafydd, 'Wiliam Bodwrda (1593–1660)', *NLWJ*, 19/1 (1975), 88–102.

Ifans, Dafydd and R. L. Thomson, 'Edward Lhuyd's *Geirieu Manaweg*', *SC*, 14/15 (1980), 129–67.

Jahn, Melvin E., 'A Note on the Editions of Edward Lhwyd's *Lithophylacii Britannici Ichnographia*', *JSBNS*, 6 (1972), 86–97.

Jahn, Melvin E., 'The Old Ashmolean Museum and the Lhwyd Collections', *JSBNS*, 4 (1966), 244–8.

January-McCann, James, 'Bywyd newydd i'r *Parochialia*', *Y Naturiaethwr*, 3(14) (2020), 8–10.

Jeffers, R. H., 'A Further Note on Edward Morgan and the Westminster Physic Garden', *Procs. Linnean Soc. London*, 168/1–2 (1957), 96–101.

Jeffers, R. H., 'Edward Morgan and the Westminster Physic Garden', *Procs. Linnean Soc. London*, 164/2 (1955), 102–33.

Jenkins, J. Geraint, *The Inshore Fishermen of Wales* (Cardiff, 1991).

Jenkins, J. Philip, 'From Edward Lhuyd to Iolo Morganwg: The Death and Rebirth of Glamorgan Antiquarianism in the Eighteenth Century', *Morgannwg*, 23 (1979), 29–47.

Jenkins, Philip, *The Making of a Ruling Class: The Glamorganshire Gentry, 1640–1790* (Cambridge, 1983).

Jenkins, R. T. and Helen Ramage, *A History of the Honourable Society of Cymmrodorion and of the Gwyneddigion and Cymreigyddion Societies*, *Y Cymmrodor*, 50 (1951).

Jessop, L., 'The Club at the Temple Coffee House – Facts and Supposition', *ANH*, 16/3 (1989), 263–74.

Johnson, Thomas, *Mercurii botanicii pars altera* (London, 1641).

Jones, Dewi, *Tywysyddion Eryri ynghyd â Nodiadau ar Lysieuaeth yr Ardal*, Llyfrau Llafar Gwlad, 25 (Llanrwst, 1993).

Jones, E. Gwynne, 'The Family Papers of Owen and Stanley of Penrhos, Holyhead', *BBCS*, 18/2 (1957), 99–115.

Jones, Stephen W., 'Thomas Francis of Montgomery (?1621–1700)', *Journal of the Montgomeryshire Genealogical Society*, 59 (2014), 25–34.

Jones, Thomas, 'Llythyr gan William Jones at Edward Lhuyd', *BBCS*, 22/3 (1967), 239–43.

Josten, C. H., 'Elias Ashmole, F.R.S. (1617–1692)', *Notes and Records of the Royal Society of London*, 15 (July 1960), 221–30.

Keller, Vera and Leigh T. I. Penman, 'From the Archives of Scientific Diplomacy: Science and the shared interests of Samuel Hartlib's London and Frederick Clodius's Gottorf', *Isis*, 106/1 (2015), 17–42.

Kendrick, T. D., *British Antiquity* (London, 1950).

Kennedy, Alasdair, 'In Search of the "True Prospect": Making and Knowing the Giant's Causeway as a Field Site in the Seventeenth Century', *The British Journal for the History of Science*, 41/1 (2008), 19–41.

Kenyon, J. R., 'William Baxter and Edward Lhuyd's *Archaeologia Britannica*', *BBCS*, 34 (1987), 118–20.

Kilburn, Matthew, 'The Fell Legacy 1686–1755', in *The History of Oxford University Press*, vol. I: *Beginnings to 1780*, ed. Ian Gadd (Oxford, 2013), pp. 107–37.

Knight, Jeremy, 'Welsh Stones and Oxford Scholars: Three Rediscoveries', in *The Afterlife of Inscriptions*, ed. Alison Cooley (London, 2000), pp. 91–101.

Le Bris, Daniel, 'Les études linguistiques d'Edward Lhuyd en Bretagne en 1701', *La Bretagne linguistique*, 14 (2009), 175–93.

Leibniz, G. W., *Collectanea etymologica, illvstrationi lingvarvm, veteris Celticæ ... inservientia* (Hanover, 1717).

Leland, John, *The Itinerary of John Leland ...* ed. Thomas Hearne, 9 vols (Oxford, 1710–12).

Leoni, Simona Boscani, 'Queries and Questionnaires: Collecting Local and Popular Knowledge in 17th and 18th Century Europe', in *Wissenchaftsgeschichte und Geschichte des Wissen im Dialog = Connecting Science and Knowledge*, ed. Kaspar von Greyerz, Silvia Flubacher and Philip Senn (Göttingen, 2013), pp. 187–210.

Levine, Joseph M., *Dr. Woodward's Shield: History, Science, and Satire in Augustan England* (Ithaca and London, 1977).

Levine, Joseph M., *The Battle of the Books: History and Literature in the Augustan Age* (Ithaca and London, 1991).

Levitin, Dmitry, 'Halley and the Eternity of the World Revisited', *Notes & Records of the Royal Society*, 67 (2013), 315–29.

Lewis, Barry, 'A Possible Provenance for the Old Cornish Vocabulary?', *Cambrian Medieval Celtic Studies*, 73 (2017), 1–14.

Lhwyd, Edward, 'A Note concerning an Extraordinary Hail in Monmouthshire', *PT*, 19 (1695–7), 579–80.

Lhwyd, Edward, *Archæologia Britannica, Texts & Translations*, ed. Dewi W. Evans and Brynley F. Roberts, Celtic Studies Publications, 10 (Aberystwyth, 2009).

Lister, Martin, *A Journey to Paris in the Year 1698* (London, 1698).

[Lloyd, Edward], 'The true Narrative of Capt. Edward Lloyd's actions & sufferings in & for his Matie's service with ye just state off his Arreares', *Bye-Gones* (1887), 413–14.

Lloyd, J. Y. W., *The History of the Princes, the Lords Marcher, and the Ancient Nobility of Powys Fadog*, 6 vols (London, 1881–7).

Lloyd, Nesta, 'Meredith Lloyd', *JWBS*, XI/3–4 (1975–6), 133–92.

Lloyd, Nesta, 'The correspondence of Edward Lhwyd and Richard Mostyn', *Flintshire Historical Society Publications*, 25 (1971–2), 32–61.

Lockton, Alex and Sarah Whild, 'The Botanical Exploration of the Oswestry District', *Shropshire Botanical Society Newsletter*, 9 (Autumn 2003), 8–12.

Lohan, Máire, 'Ceremonial Monuments in Moytura, Co. Mayo', *Journal of the Galway Archaeological and Historical Society*, 51 (1999), 77–108.

Lucey, John, 'Irish Spurge (*Euphorbia hyberna*) in England: native or naturalised?', *British & Irish Botany*, 1/3 (2019), 243–9.

Maber, R. G., 'Celtic Prosody in Late Cornish: The *englyn* "An lavar kôth yu lavar guîr"', *BBCS*, 28/4 (1980), 593–8.

MacGregor, Arthur, *Ark to Ashmolean: The Story of the Tradescants, Ashmole and the Ashmolean Museum* (Oxford, 1983).

MacGregor, Arthur, 'Edward Lhuyd, Museum Keeper', *WHR*, 25/1 (2010), 51–74.

MacGregor, Arthur, *The Ashmolean Museum: A Brief History of the Museum and its Collections* (Oxford and London, 2001).

MacGregor, Arthur, *Tradescant's Rarities: Essays on the Foundation of the Ashmolean Museum, 1683, with a Catalogue of the Surviving Early Collections* (Oxford, 1983), available online at *http://etheses.dur.ac.uk/10281/1/10281_7075.PDF?UkUDh:CyT.*

MacGregor, Arthur and Moira Hook, *Manuscript Catalogues of the Early Museum Collections (Part 2): The Vice-Chancellor's Consolidated Catalogue, 1695*, BAR International Series, 1569 (Oxford, 2006).

MacGregor, Arthur, Melanie Mendonça and Julia White, *Manuscript Catalogues of the Early Museum Collections, 1683–1886*, Part 1, BAR International Series, 907 (Oxford, 2000).

MacGregor, Arthur and A. Turner, 'The Ashmolean Museum', in *The History of the University of Oxford*, V: *The Eighteenth Century*, ed. L. S. Sutherland and L. G. Mitchell (Oxford, 1986), pp. 639–58.

McAteer, Claire and Matthew Parkes, *The Geological Heritage of Sligo: An Audit of County Geological Sites in Sligo*, Geological Survey of Ireland (2004), available online at *https://secure.dccae.gov.ie/GSI_DOWNLOAD/Geoheritage/Reports/Sligo_Audit.pdf.*

McGuinness, David, 'Edward Lhuyd's Contribution to the Study of Irish Megalithic Tombs', *Journal of the Royal Society of Antiquaries of Ireland*, 126 (1996), 62–85.

McNamara, Ken, *Dragon's Teeth and Thunderstones: The Quest for the Meaning of Fossils* (London, 2020).

Mandelbrote, Scott, 'The Publication and Illustration of Robert Morison's *Plantarum historiae universalis Oxoniensis*', *Huntington Library Quarterly*, 78/2 (2015), 349–79.

Marrinan, Lauren, and Tim Rich, *101 Rare Plants of Wales* (Llanelli, 2019).

Matheson, Colin, *Wales and the Sea Fisheries* (Cardiff, 1929).

Mendyk, S., 'Robert Plot: Britain's "Genial father of County Natural Histories"', *Notes and Records of the Royal Society of London*, 39/2 (April 1985), 159–77.

Mendyk, Stan A. E., '*Speculum Britanniae': Regional Study, Antiquarianism, and Science in Britain to 1700* (Toronto and London, 1989).

Mitchell, M. E., 'Irish Botany in the Seventeenth Century'. *Proceedings of the Royal Irish Academy, Section B: Biological, Geological, and Chemical Science*, 75 (1975), 275–84.

Moon, B. C. and A. M. Kirton, *Ichthyosaurs of the British Middle and Upper Jurassic. Part 1: Ophthalmosaurus*, Monograph of the Palaeontographical Society, 170 (647) (London, 2016), pp. 1–84.

Morgan, P. J. T., 'The Abbé Pezron and the Celts', *THSC* (1965), 286–95.

Morris, Owen, *The 'Chymic Bookes' of Sir Owen Wynne of Gwydir; An Annotated Catalogue*, Libri Pertinentes, No. 4 (Cambridge, 1997).

Morris-Jones, John, 'Edward Lhwyd', *Y Traethodydd*, 49 (1893), 465–75.

Morris-Jones, John, *Edward Lhwyd: Inaugural Address Delivered at the University College of North Wales, Session 1893–94* (Caernarfon, 1894).

Morse, Michael A., *How the Celts Came to Britain: Druids, Ancient Skulls and the Birth of Archaeology* (Stroud, 2005).

Moyle, Walter, *The Works of Walter Moyle* ... 2 vols (London 1726).

Nelson, E. C., 'A History, mainly Nomenclatural, of St. Dabeoc's Heath', *Watsonia*, 23 (2000), 47–58.

Nelson, E. C., *Sea Beans and Nickar Nuts: A Handbook of Exotic Seeds and Fruits Stranded on Beaches in North-Western Europe*, B.S.B.I. Handbooks for Physical Identification, 10 (London, 2000).

Newman, John and Nicholas Pevsner, *Shropshire*, The Buildings of England (New Haven and London, 2006).

Nicolson, William, *The London Diaries of William Nicolson, Bishop of Carlisle, 1702–1718*, ed. Clyve Jones and Geoffrey Holmes (Oxford, 1985).

North, F. J., *Coal, and the Coalfields in Wales* (Cardiff, 1926).

Oakley, Richard R, *A History of Oswestry School* (London, [1964]).

Oates, J. C. T., *Cambridge University Library: A History. From the Beginnings to the Copyright Act of Queen Anne* (Cambridge, 1986).

Oates, J. C. T., 'Notes on the Later History of the Oldest Manuscript of Welsh Poetry: The Cambridge Juvencus', *Cambridge Medieval Celtic Studies*, 3 (1982), 81–7.

O'Sullivan, Anne and William O'Sullivan, 'Edward Lhuyd's Collection of Irish Manuscripts', *THSC* (1962), 57–76.

Oswald, Philip H., 'Edward Morgan and his *Hortus Siccus* as a Source of Early Records of Welsh Plants', *Archives of Natural History* (forthcoming).

Oswestry Ramblers, *More Favourite Walks around Oswestry and the Borders* (Llanrhaeadr ym Mochnant, 2011).

Ovenden, Richard, 'The Learned Press: Printing for the University', in *The History of Oxford University Press*, vol. I: *Beginnings to 1780*, ed. Ian Gadd (Oxford, 2013), pp. 279–92.

Ovenell, R. F., *The Ashmolean Museum, 1683–1894* (Oxford, 1986).

Owen, G. Dyfnallt, *Wales in the Reign of James I*, Royal Historical Society Studies in History, 53 (Woodbridge, 1998).

Owen, Trefor M., *Welsh Folk Customs* (Cardiff, 1968).

Owens, R. M., *Trilobites in Wales* (Cardiff, 1984).

Oxford in 1710 from the Travels of Zacharias Conrad von Uffenbach, trans. and ed. W. H. Quarrell and W. J. C. Quarrell (Oxford, 1928).

P., L., *Two Essays sent in a Letter from Oxford to a Nobleman in London* (London, 1695).

Padel, Oliver, 'The Nature and Date of the Old Cornish Vocabulary', *ZCP*, 61 (2014), 173–99.

Parochialia, Being a Summary of Answers to "Parochial Queries in Order to a Geographical Dictionary, etc., of Wales" issued by Edward Lhuyd, ed. Rupert H. Morris, *Archaeologia Cambrensis*, Supplements (London, 1909–11).

Parry, Graham, *The Trophies of Time: English Antiquarians of the Seventeenth Century* (Oxford, 1995).

Parry-Jones, J., 'The Story of Oswestry Castle', *Trans. Shropshire Archaeological and Natural History Society*, 2nd series, 6 (1894), 107–73.

Pearman, David, *The Discovery of the Native Flora of Britain and Ireland* (Bristol, 2017).

Pepys, Samuel, *The Diary of Samuel Pepys. A New and Complete Transcription*, ed. Robert Latham and William Matthews, 11 vols (London, 1970–83).

Phillips, John Roland, *Memoirs of the Civil War in Wales and the Marches, 1642–1649*, 2 vols (London, 1874).

Piggott, Stuart, *Ruins in a Landscape: Essays in Antiquarianism* (Edinburgh, 1976).

Plot, Robert, *The Natural History of Oxford-Shire* (Oxford, 1677).

Pool, P. A. S., 'Cornish Drawings by Edward Lhuyd in the British Museum', *Cornish Archaeology*, 16 (1977), 139–42.

Poole, William, *The World Makers: Scientists of the Restoration and the Search for the Origins of the Earth* (Oxford, 2010).

Porter, Roy, *The Making of Geology: Earth Science in Britain 1660–1815* (Cambridge, 1977).

Powell, Philip, *The Geology of Oxfordshire* (Wimborne, 2005).

Pryce, Arthur Ivor, *The Diocese of Bangor during Three Centuries ... being a Digest of the Registers of the Bishops* (Cardiff, 1929).

Pryce, William, *Archæologia Cornu-Britannica* (Sherborne, 1790).

Pryce-Jones, John, 'Oswestry Corporation Records – the Bailiffs from Medieval Times to 1673', *Shropshire History and Archaeology, Transactions of the Shropshire Archaeological and Historical Society*, 76 (2001), 30–9.

Rappaport, Rhoda, 'Questions of Evidence: An Anonymous Tract Attributed to John Toland', *Journal of the History of Ideas*, 58 (1997), 339–48.

Rappaport, Rhoda, *When Geologists were Historians, 1665–1750* (Ithaca and London, 1997).

Raven, Charles E., *English Naturalists from Neckam to Ray: A Study of the Making of the Modern World* (Cambridge, 1947).

Raven, Charles E., *John Ray, Naturalist: His Life and Works* (Cambridge, 1950).

Ray, John, *A Collection of English Words Not Generally Used* ... 2nd edn (London, 1691).

[Ray, John], *Further Correspondence of John Ray*, ed. R. T. Gunther (London, 1928).

Ray, John, *Miscellaneous Discourses Concerning the Dissolution and Changes of the World* ... (London, 1692).

Ray, John, *Synopsis Methodica Stirpium Britannicarum* (London, 1690).

Ray, John, *Synopsis Methodica Stirpium Britannicarum* ... 3rd edn (London 1724) facsimile with an introduction by William T. Stearn (London, 1973).

[Ray, John], *The Correspondence of John Ray: Consisting of Selections from the Philosophical Letters published by Dr. Derham, and Original Letters* ... ed. Edwin Lankester (London, 1848).

Raye, Lee, 'Robert Sibbald's *Scotia Illustrata* (1684): A Faunal Baseline for Britain', *Notes and Records*, 72 (2018), 383–405.

Recusants in the Exchequer Pipe Rolls 1581–1592 ... ed. Timothy J. McCann, Catholic Record Society Publications (Records Series), vol. 71 (n.p., 1986).

Redknap, M. and J. M. Lewis, *A Corpus of Early Medieval Inscribed Stones and Stone Sculpture in Wales*, vol.1: *South Wales* (Cardiff, 2007).

Rees, Eiluned and Gwyn Walters, 'The Dispersion of the Manuscripts of Edward Lhuyd', *WHR*, 7/2 (1974), 148–78.

Riley, Margaret, 'The Club at the Temple Coffee House Revisited', *ANH*, 33/1 (2006), 90–100.

Ritchie, J. C., 'Vaccinium vitis-ideae L.', *Journal of Ecology*, 43/2 (1955), 701–8.

Roberts, Brynley F., 'A Note on the Ashmolean Collection of Letters addressed to Edward Lhuyd', *WHR*, 7/2 (1975), 179–85.

Roberts, Brynley F., 'An Early Edward Lhwyd Glossary', *SC*, 50 (2016), 151–62.

Roberts, Brynley F., *'Anwir Anwedhys y mae yn i Ysgrivennv Ymma': Rhai o Ymylnodau Edward Lhwyd* (Aberystwyth, 2009).

Roberts, Brynley F., *Brut Tysilio* (Swansea 1980).

Roberts, Brynley F., 'Cyhoeddiadau Edward Lhwyd', *Y Llyfr yng Nghymru/ Welsh Book Studies*, 1 (1998), 21–58.

Roberts, Brynley F., 'Cymraeg Edward Lhwyd', *Y Traethodydd*, 158 (2003), 211–28.

Roberts, Brynley F., 'Dialedd Taffy (*Hoglandiae Descriptio* Thomas Richards)', *Ysgrifau Beirniadol*, 25 (1999), 61–78.

Roberts, Brynley F., 'Edward Lhuyd and Celtic Linguistics', in *Proceedings of the Seventh International Congress of Celtic Studies*, ed. D. Ellis Evans, John G. Griffith and E. M. Jope (Oxford, 1986), pp. 1–9.

Roberts, Brynley F., 'Edward Lhuyd a'r Bywyd Diwylliannol Cymreig', *Cof Cenedl*, 18 (2003), 37–69.

Roberts, Brynley F., 'Edward Lhuyd – Welshman', *Nature in Wales*, NS 2/1–2 (1984). An abbreviated translation of his 'Edward Lhuyd y Cymro' (below).

Roberts, Brynley F., 'Edward Lhuyd y Cymro', *NLWJ*, 24 (1985), 63–83.

Roberts, Brynley F., 'Edward Lhwyd a Cheredigion', *Ceredigion*, 16 (2009), 49–69.

Roberts, Brynley F., 'Edward Lhwyd (*c.*1660–1709): Folklorist', *Folklore*, 120 (2009), 36–56.

Roberts, Brynley F., 'Edward Lhwyd in Carmarthenshire, *The Carmarthenshire Antiquary*, 46 (2010), 24–43.

Roberts, Brynley F., 'Edward Lhwyd in Cornwall', *SC*, 53 (2019), 133–53.

Roberts, Brynley F., 'Edward Lhuyd's Collection of Printed Books', *The Bodleian Library Record*, 10 (1979), 112–27.

Roberts, Brynley F., 'Edward Lhuyd's Debts', *BBCS*, 26/3 (1975), 353–9.

Roberts, Brynley F., 'Edward Lhuyd's Protégés', *THSC*, NS, 14 (2008), 21–57.

Roberts, Brynley F., 'Etifeddion Edward Lhuyd', *WHR*, 25/1 (2010), 97–119.

Roberts, Brynley F., 'Lloyd – Lhuyd – Lhwyd', *Y Traethodydd*, 151 (1996), 180–3.

Roberts, Brynley F., 'Llythyrau John Lloyd at Edward Lhuyd', *NLWJ*, 17/1 (1971), 88–114; 17/2 (1971), 183–206.

Roberts, Brynley F., '"Memoirs of Edward Lhwyd, Antiquary" and Nicholas Owen's *British Remains*, 1777', *NLWJ*, 19/1 (1975), 67–87.

Roberts, Brynley F., 'Pedwar Portread o Edward Lhuyd', *NLWJ*, 21/1 (1979), 40–2.

Roberts, Brynley F., 'Richard Ellis, M.A.: Edward Lhuyd and the Cymmrodorion', *THSC* (1977), 131–72 (pp. 166–72).

Roberts, Brynley F., 'Scholarly publishing 1820–1922', in *A Nation and its Books: A History of the Book in Wales*, ed. Philip Henry Jones and Eiluned Rees (Aberystwyth, 1998), pp. 221–35.

Roberts, Brynley F., 'The Discovery of Old Welsh', *Historiographia Linguistica*, 26 (1999), 10–21.

Roberts, Brynley F., 'Translating Old Welsh: The First Attempts', *ZCP*, 49–50 (1997), 760–77.

Robinson, Tancred, *A Letter sent to Mr. William Wotton … Concerning Some Late Remarks &c. Written by John Harris* ([London, 1697]).

Roderick O'Flaherty's Letters 1696–1709: To William Molyneux, Edward Lhwyd, and Samuel Molyneux, ed. with notes and an introduction by Richard Sharpe (Dublin, 2013).

Roebuck, Thomas, 'Edmund Gibson's 1695 *Britannia* and Late-Seventeenth-Century British Antiquarian Scholarship', *Erudition and the Republic of Letters*, 5 (2020), 427–81.

Roos, Anna Marie, 'Only meer love to Learning: A Rediscovered Travel Diary of Naturalist and Collector James Petiver (*c.*1665–1718)', *Journal of the History of Collections*, 29 (2017), 381–94.

Roos, Anna Marie, *Web of Nature: Martin Lister (1639–1712), the First Arachnologist* (Leiden and Boston, 2011).

Roos, Anna Marie and Edwin D. Rose, 'Lives and Afterlives of the *Lithophylacii Britannici ichnographia* (1699), the First Illustrated Field Guide to English Fossils', *Nuncius*, 33 (2018), 505–36.

Rudwick, Martin J. S., *Earth's Deep History: How it was Discovered and Why it Matters* (Chicago and London, 2014).

Rudwick, Martin J. S., *The Meaning of Fossils: Episodes in the History of Palaeontology*, 2nd edn (Chicago and London, 1976).

Ruscoe, Anthony, *Landed Estates and the Gentry: The Country South of Oswestry* (Ormskirk, 2006).

Scheuchzer, Johann Jakob, ΟΥΡΕΣΙΦΟΙΤΕΣ *sive itinera alpina tria* (London, 1708).

Seeber, Karin, 'Jacob Bobart (1596–1680); First Keeper of the Oxford Physic Garden', *Garden History*, 41/2 (2013), 278–84.

Sendino, Consuelo, 'The Hans Sloane Fossil Collection at the Natural History Museum, London', *Deposits Mag*, 47 (2016), 13–17.

Sharpe, Richard, 'Lachlan Campbell's Letters to Edward Lhwyd 1704–7', *Scottish Gaelic Studies*, 29 (2013), 244–81.

Shaw, Francis, 'The Background to *Grammatica Celtica*', *Celtica*, 3 (1956), 1–16.

Shropshire Parish Registers, Diocese of St Asaph, vol. IV: *The Register of Oswestry*, vol. I (n.p., 1909).

Shropshire Parish Registers, Diocese of St Asaph, vol. V: *Register of Oswestry*, vols II–III, (n.p., 1912).

Siddons, Michael Powell, *The Development of Welsh Heraldry*, 3 vols (Aberystwyth, 1991–3).

Siddons, Michael Powell, *Welsh Pedigree Rolls* (Aberystwyth, 1996).

Simcock, A. V., *The Ashmolean Museum and Oxford Science, 1683–1983* (Oxford, 1984).

The Sloane Herbarium: An Annotated List of the Horti Sicci composing it, with Biographical Accounts of the Principal Contributors based on Records compiled by … James Britten, rev. and ed. J. E. Dandy (London, 1958).

Smiles, Sam, *The Image of Antiquity: Ancient Britain and the Romantic Imagination* (New Haven and London, 1994).

Smith, Llinos, 'Oswestry', in R. A. Griffiths (ed.), *Boroughs of Mediaeval Wales* (Cardiff, 1978), pp. 218–42.

Smith, M. G., *'Fighting Joshua': A Study of the Career of Sir Jonathan Trelawny, Bart., 1650–1721, Bishop of Exeter, and Winchester* (Redruth, 1985).

Stace, Clive, *New Flora of the British Isles*, 4th edn (Middlewood Green, 2019).

Stamper, Paul, *Historic Parks and Gardens of Shropshire* (n.p., 1996).

Stamper, Paul, 'Where Samson Pruned Roses', *British Archaeology*, 53 (June 2000), 42.

Stokes, Evelyn, 'Volcanic Studies by Members of the Royal Society of London 1665–1780', *Earth Science Journal*, 5/2 (1971), 46–67.

Stout, Geraldine and Matthew Stout, *Newgrange* (Cork, 2008).

Thick, Malcolm, 'Garden Seeds in England before the late eighteenth century – II, The Trade in Seeds to 1760', *Agricultural History Review*, 38/2 (1990), 105–16.

Thick, Malcolm, 'The Sale of Produce from Non-commercial Gardens in late medieval and early modern England', *Agricultural History Review*, 66/1 (2018), 1–17.

Thomas, D. R., *The History of the Diocese of St. Asaph, General, Cathedral and Parochial*, new enlarged and illustrated edn, 3 vols (Oswestry, 1908–13).

Thomson, R. L., 'Edward Lhuyd in the Isle of Man?', in *Celtic Studies: Essays in Memory of Angus Matheson, 1912–1962*, ed. James Carney and David Greene (London, 1969), pp. 170–82.

Tite, Colin G. C., *The Manuscript Library of Sir Robert Cotton*, The Panizzi Lectures, 1993 (London, 1994).

Turner, A. J., 'A Forgotten Naturalist of the Seventeenth Century: William Cole of Bristol and his Collections', *ANH*, 11/1 (1982), 27–41.

Victoria History of the Counties of England: A History of Shropshire, vol. 2, ed. A. T. Gaydon (London, 1973).

Victoria History of the Counties of England: A History of Shropshire, vol. 3, ed. C. R. Elrington (Oxford, 1979).

Waddell, Brodie, 'The Politics of Economic Distress in the Aftermath of the Glorious Revolution, 1689–1702', *English Historical Review*, 130 (2015), 318–51.

Walters, Gwyn and Frank Emery, 'Edward Lhuyd, Edmund Gibson, and the Printing of Camden's *Britannia*, 1695', *The Library*, 5th series, 32 (1977), 109–37.

Watkin, Isaac, *Oswestry, with an Account of its Old Houses, Shops, Etc., and some of their Occupants* (London and Oswestry, 1920).

Williams, Derek R., *Edward Lhuyd (1660–1709): A Shropshire Welshman* (Oswestry, 2009).

Williams, Derek R., *Prying into Every Hole and Corner: Edward Lhuyd in Cornwall in 1700* (Trewirgie, 1993).

Williams, G. J., 'Edward Lhuyd', *Llên Cymru*, 6 (1961), 122–37.

Williams, G. J., 'Edward Lhuyd a Thraddodiad Ysgolheigaidd Sir Ddinbych', *Denbighshire Historical Society Transactions*, 11 (1962), 1–21.

Williams, G. J., *Edward Lhuyd ac Iolo Morganwg: Agweddau ar Hanes Astudiaethau Gwerin yng Nghymru* (Cardiff, 1964).

Williams, G. J., *Traddodiad Llenyddol Morgannwg* (Cardiff, 1948).

Williams, Ifor, *The Beginnings of Welsh Poetry: Studies*, ed. Rachel Bromwich (Cardiff, 1972).

Williams, Robert J. P., John S. Rowlinson and Allan Chapman (eds), *Chemistry at Oxford: A History from 1600 to 2005* (Cambridge, 2009).

Williams, W. R., *The History of the Great Sessions in Wales 1542–1830 …* (Brecknock, 1899).

Worton, Jonathan, 'The Royalist and Parliamentarian War Effort in Shropshire during the First and Second English Civil Wars, 1642–1648' (unpublished doctoral thesis, University of Chester, 2015), at *https:// chesterrep.openrepository.com/bitstream/handle/10034/612966/Main%20article. pdf?isAllowed=y&sequence=1*.

Yale, Elizabeth, *Sociable Knowledge: Natural History and the Nation in Early Modern Britain* (Philadelphia, 2016).

INDEX

Wherever possible plants are entered under their current binomial names. Animals are entered under their common English names.